PROCESS SAFETY IN UPSTREAM OIL AND GAS

This book is one of a series of process safety guidelines and concept books published by the Center for Chemical Process Safety (CCPS). Please go to www.wiley.com/go/ccps for a full list of titles in this series.

This concept book is issued jointly with the Society of Petroleum Engineers (SPE). SPE publications can be found at SPE.org.

PROCESS SAFETY IN UPSTREAM OIL AND GAS

CENTER FOR CHEMICAL PROCESS SAFETY
of the
AMERICAN INSTITUTE OF CHEMICAL ENGINEERS
120 Wall St, 23rd Floor • New York, NY 10005

Registered Office
John Wiley & Sons, Inc., 111 River Street, Hoboken, NJ 07030, USA

Editorial Office
111 River Street, Hoboken, NJ 07030, USA

For details of our global editorial offices, customer services, and more information about Wiley products visit us at www.wiley.com.

Wiley also publishes its books in a variety of electronic formats and by print-on-demand. Some content that appears in standard print versions of this book may not be available in other formats.

Library of Congress Cataloging-in-Publication Data
Names: American Institute of Chemical Engineers. Center for Chemical Process Safety, author.
Title: Process safety in upstream oil and gas / American Institute of Chemical Engineers.
Other titles: Process safety in upstream oil and gas
Description: Hoboken, NJ, USA : Wiley, 2021. | Includes bibliographical references and index.
Identifiers: LCCN 2021003994 (print) | LCCN 2021003995 (ebook) | ISBN 9781119620044 (hardback) | ISBN 9781119620051 (adobe pdf) | ISBN 9781119620143 (epub)
Subjects: LCSH: Oil fields–Safety measures. | Gas wells–Safety measures. | Petroleum industry and trade–Safety measures. | Gas industry–Accidents.
Classification: LCC TN871.215 .P76 2021 (print) | LCC TN871.215 (ebook) | DDC 622/.8–dc23
LC record available at https://lccn.loc.gov/2021003994
LC ebook record available at https://lccn.loc.gov/2021003995

Cover Design: Wiley
Cover Images: © grandriver/Getty Images; © HeliRy/Getty Images

CONTENTS

LIST OF TABLES

LIST OF FIGURES

ACRONYMS AND ABBREVIATIONS

ADNOC	Abu Dhabi National Oil Company
AEGL	Acute Exposure Guideline Level
AIChE	American Institute of Chemical Engineers
ALARP	As Low as Reasonably Practicable
ANP	National Agency of Oil, Gas and Biofuels
ANSI	American National Standards Institute
API	American Petroleum Institute
ASME	American Society of Mechanical Engineers
BLEVE	Boiling Liquid Expanding Vapor Explosion
BSEE	Bureau of Safety and Environmental Enforcement
BOD	Basis of Design
BOP	Blowout Preventer
CCF	Common Cause Failure
CCPS	Center for Chemical Process Safety (of AIChE)
CCTV	Closed Circuit Television
CFD	Computational Fluid Dynamics
C-NLOPB	Canada-Newfoundland & Labrador Offshore Petroleum Board
COMAH	Control of Major Accident Hazards (UK Regulation incorporating the EU Seveso Directive requirements)
COS	Center for Offshore Safety
CPQRA	Chemical Process Quantitative Risk Assessment
CRA	Concept Risk Analysis
CSB	Chemical Safety Board (US)
DDT	Deflagration to Detonation Transition
DOE	Department of Energy (US)
DWOP	Drill Well on Paper
EER	Escape Evacuation and Rescue
EPA	Environmental Protection Agency (US)
ERPG	Emergency Response Planning Guideline
ESD	Emergency Shutdown
ESDV	Emergency Shutdown Valve
ETA	Event Tree Analysis
EU	European Union
FABIG	Fire and Blast Information Group
FAR	Fatal Accident Rate
FAT	Factory Acceptant Test
FEED	Front End Engineering Design
FEL	Front End Loading
FERC	Federal Energy Regulatory Commission

FHA	Fire Hazard Analysis
FLNG	Floating Liquefied Natural Gas
FMEA	Failure Modes and Effects Analysis
FMECA	Failure Modes, Effects and Criticality Analysis
FPSO	Floating Production Storage and Offloading
FPU	Floating Production Unit
FTA	Fault Tree Analysis
GOM	Gulf of Mexico
GPS	Global Positioning System
HAZID	Hazard Identification Study
HAZOP	Hazard and Operability Study
HF	Hydrogen Fluoride
HIRA	Hazard Identification and Risk Analysis
HOF	Human and Organizational Factors
HP	High Pressure
HPHT	High Pressure High Temperature
HSE	Health, Safety and Environment
HSE	Health and Safety Executive (UK)
HSL	Health and Safety Laboratory (UK)
IADC	International Association of Drilling Contractors
ICRARD	International Committee on Regulatory Authority Research and Development
IEC	International Electrotechnical Commission
IOGP	International Association of Oil & Gas Producers
IP	Intermediate Pressure
IPL	Independent Protection Layer
ISD	Inherently Safety Design
ISM	International Safety Management
ISO	International Standards Organization
JSA	Job Safety Analysis
KPI	Key Performance Indicator
LMRP	Lower Marine Riser Package
LNG	Liquefied Natural Gas
LOPA	Layer of Protection Analysis
LOPC	Loss of Primary Containment
LP	Low Pressure
LPG	Liquefied Petroleum Gas
LSIR	Location Specific Individual Risk
MAE	Major Accident Event
MDEA	Methyl diethanolamine
MEA	Ethanolamine
MKO	Mary Kay O'Connor Process Safety Center
MOC	Management of Change
MOOC	Management of Organizational Change
MODU	Mobile Offshore Drilling Unit
NACE	National Association of Corrosion Engineers

NASA	National Aeronautics and Space Administration (US)
NFPA	National Fire Protection Association
NLOPB	National Offshore Petroleum Safety and Environmental Management Authority
NOPSEMA	National Offshore Petroleum Safety and Environmental Management Authority (Australia)
NORSOK	Norwegian Oil Industry Standards (Norsk Sokkels Konkuranseposisjon)
NORM	Naturally Occurring Radioactive Materials
OCA	Offsite Consequence Analysis
OCS	Outer Continental Shelf
OESI	Ocean Energy Safety Institute
OSHA	Occupational Safety and Health Administration (US)
PDCA	Plan – Do – Check – Act
PFD	Process Flow Diagram
PFP	Passive Fire Protection
PHA	Process Hazard Analysis
P&ID	Piping and Instrumentation Diagram
PHMSA	Pipeline and Hazardous Materials Safety Administration (US)
PLL	Potential Loss of Life
PMBOK	Project Managers Book of Knowledge
PPE	Personal Protective Equipment
PPG	Pounds per Gallon
PPM	Parts per Million
PSA	Petroleum Safety Authority (Norway)
PSM	Process Safety Management
PTW	Permit to Work
QA	Quality Assurance
QC	Quality Control
QRA	Quantitative Risk Assessment
RAGAGEP	Recognized and Generally Accepted Good Engineering Practice
RAM	Reliability, Availability, and Maintainability
RASCI	Responsible, Accountable, Supporting, Consulted and Informed
RBPS	Risk Based Process Safety
R&D	Research and Development
RMP	Risk Management Program
ROV	Remotely Operated Vehicles
RP	Recommended Practice
RPSEA	Research Partnership to Secure Energy for America
SAChE	Safety and Chemical Engineering Education
SCE	Safety Critical Element (also Safety or Environmental Critical Element or Equipment)
SEMS	Safety and Environmental Management Systems (US)
SIL	Safety Integrity Level (as per IEC 61508 / 61511 standards)

SIMOPS	Simultaneous Operations
SINTEF	Independent research institute (Norway)
SIS	Safety Instrumented System
SPE	Society of Petroleum Engineers
SSSV	Sub-surface Safety Valve
TAP	Technology Assessment Program
TLP	Tension Leg Platform
TSR	Temporary Safe Refuge
UFL	Upper Flammability Limit
USCG	United States Coast Guard
VCE	Vapor Cloud Explosion

GLOSSARY

Terms in this Glossary are taken from the online CCPS Glossary of Terms for Process Safety or the Schlumberger Oilfield Glossary, where available.

ALARP As Low As Reasonably Practicable – a term used to describe a target level for reducing risk that would implement risk reducing measures unless the costs of the risk reduction in time, trouble or money are grossly disproportionate to the benefit.

Barrier A control measure or grouping of control elements that on its own can prevent a threat developing into a top event (prevention barrier) or can mitigate the consequences of a top event once it has occurred (mitigation barrier). A barrier must be effective, independent, and auditable. See also **Degradation Control**. (Other possible names: **Control, Independent Protection Layer, Risk Reduction Measure**).

Blowout Uncontrolled flow of formation fluids from a well. An uncontrolled flow of formation fluids from the wellbore or into lower pressured subsurface zones (underground blowout). Uncontrolled flows cannot be contained using previously installed barriers and require specialized services intervention.

A blowout may consist of water, oil, gas or a mixture of these. Blowouts may occur during all types of well activities and are not limited to drilling operations. In some circumstances, it is possible that the well will bridge over, or seal itself with rock fragments from collapsing formations downhole.

Blowout Preventer (BOP)   A large valve or assembly of valves at the top of a well that may be closed if the drilling crew loses control of formation fluids. By closing this valve (usually operated remotely via hydraulic actuators), the drilling crew usually regains control of the reservoir, and procedures can then be initiated to increase the mud density until it is possible to open the BOP and retain pressure control of the formation. BOPs come in a variety of styles, sizes and pressure ratings. Some can effectively close over an open wellbore, some are designed to seal around tubular components in the well (drillpipe, casing or tubing) and others are fitted with hardened steel shearing surfaces that can actually cut through drillpipe. The BOP includes potentially several types of ram and annular devices to stop flows of well fluids, such as annular ram, shear ram, variable bore rams and pipe rams.

Bow Tie Model   A risk diagram showing how various threats can lead to a loss of control of a hazard and allow this unsafe condition to develop into a number of undesired consequences. The diagram can show all the barriers and degradation controls deployed.

Casing   Steel pipe cemented in place during the construction process to stabilize the wellbore. The casing forms a major structural component of the wellbore and serves several important functions: preventing the formation wall from caving into the wellbore, isolating the different formations to prevent the flow or crossflow of formation fluid, and providing a means of maintaining control of formation fluids and pressure as the well is drilled.

Cement   The material used to permanently seal annular spaces between casing and borehole walls. Cement is also used to seal formations to prevent loss of drilling fluid and for operations ranging from setting kick-off plugs to plug and abandonment.

Christmas Tree   An assembly of valves, spools and fittings connected to the wellhead used to control the flow produced by a well. Offshore the Christmas tree may be located on the seabed (wet tree) or on the surface facility (dry tree).

Completion	Completion of a well. The process by which a well is brought to its final classification - basically dry hole, producer, or injector. A dry hole is completed by plugging and abandonment. A well deemed to be producible of petroleum, or used as an injector, is completed by establishing a connection between the reservoir(s) and the surface so that fluids can be produced from or injected into the reservoir. Various methods are utilized to establish this connection, but they commonly involve the installation of some combination of borehole equipment, casing and tubing, and surface injection or production facilities.
Consequence	The undesirable result of a loss event, usually measured in health and safety effects, environmental impacts, loss of property, and business interruption costs. Another possible name: **Outcome**. The magnitude of the consequence may be described using a Risk Matrix
Conventional Reservoir	Traditionally, a conventional reservoir was defined where buoyant forces keep oil and gas sealed beneath a caprock and these can flow easily into a wellbore for production. An updated view defines conventional oil or gas as coming from formations that are "normal" or straightforward to extract product from. Extracting fossil fuels from these geological formations can be done with standard methods that can be used to economically remove the fuel from the deposit. Conventional resources tend to be easier and less expensive to produce simply because they require no specialized technologies and can utilize common methods.
Degradation Factor	A situation, condition, defect, or error that compromises the function of a main pathway barrier, through either defeating it or reducing its effectiveness. If a barrier degrades then the risks from the pathway on which it lies increase or escalate, hence the alternative name of escalation factor. (Other possible names: **Barrier Decay Mechanism, Escalation Factor, Defeating Factor**).
Degradation Control	Measures which help prevent the degradation factor impairing the barrier. They lie on the pathway connecting the degradation threat to the main pathway barrier. Degradation controls may not meet the full requirements for barrier validity. (Other possible names: **Degradation Safeguard, Defeating Factor Control, Escalation Factor Control, Escalation Factor Barrier**).

Directional Drilling	The intentional deviation of a wellbore from the path it would naturally take. Directional drilling is common because it allows drillers to place the borehole in contact with the most productive reservoir rock.
Drill String	The combination of the drill pipe, the bottom hole assembly and any other tools used to make the drill bit rotate at the bottom of the wellbore.
Drilling	The process whereby a hole is bored using a drill bit to create a well for oil and natural gas production and establish key parameters for the well.
Exploration	The initial phase in petroleum operations that includes generation of a prospect or play or both and drilling of an exploration well. Appraisal, development and production phases follow successful exploration.
Fracture Gradient	The pressure required to induce fractures in rock at a given depth. If the fracture gradient is exceeded, then some of the dense drilling mud can be lost into the formation leading to a potential loss of hydrostatic head. If the pressure exerted by the hydrostatic head falls below the local pore pressure in the reservoir zone, then hydrocarbons can flow into the well.
Front End Loading	The process for conceptual development of projects. This involves developing sufficient strategic information with which owners can address risk and make decisions to commit resources in order to maximize the potential for success. (Other possible names: **pre-project planning, front-end engineering design, feasibility analysis, early project planning**).
Hazard	An operation, activity or material with the potential to cause harm to people, property, the environment or business; or simply, a potential source of harm.
Hazard Identification and Risk Analysis	A collective term that encompasses all activities involved in identifying hazards and evaluating risk at facilities, throughout their life cycle, to make certain that risks to employees, the public, or the environment are consistently controlled within the organization's risk tolerance.
Incident	An event, or series of events, resulting in one or more undesirable consequences, such as harm to people, damage to the environment, or asset/business losses. Such events include fires, explosions, releases of toxic or otherwise harmful substances, and so forth.

Kick	An unexpected and unwanted influx of fluid into the wellbore during drilling operations.
LOPA	Layer of Protection Analysis. An approach that analyzes one incident scenario (cause-consequence pair) at a time, using predefined values for the initiating event frequency, independent protection layer failure probabilities, and consequence severity, in order to compare a scenario risk estimate to risk criteria for determining where additional risk reduction or more detailed analysis is needed.
Loss of Primary Containment	An unplanned or uncontrolled release of material from primary containment, including non-toxic and non-flammable materials (e.g., steam, hot condensate, nitrogen, compressed CO_2 or compressed air).
Loss of Well Control	Uncontrolled flow of formation or other fluids. The flow may be to an exposed formation (an underground blowout) or at the surface (a surface blowout).
Major Accident Event (MAE)	A hazardous event that results in one or more fatalities or severe injuries; or extensive damage to structure, installation or plant; or large-scale, severe and / or persistent impact on the environment. In bow tie modeling, MAEs are outcomes of the top event. (Other possible names: **major accident, major incident**).
Management of Change	A management system to identify, review, and approve all modifications to equipment, procedures, raw materials, and processing conditions, other than replacement in kind, prior to implementation to help ensure that changes to processes are properly analyzed (for example, for potential adverse impacts), documented, and communicated to employees affected.
Mitigation Barrier	A barrier located on the right-hand side of a bow tie diagram lying between the top event and a consequence. It might only reduce a consequence, not necessarily terminate the sequence before the consequence occurs (Other possible names: **Reactive Barrier, Recovery Measure**).
Mud Column	The height of liquid drilling mud measured in feet (meters) from the bottom to the top of a borehole, either while being circulated during drilling operations or when the drill string is not in the hole. (Other possible name: **Fluid Column**).

Overbalance	The amount of pressure (or force per unit area) in the wellbore that exceeds the pressure of fluids in the formation. This excess pressure is needed to prevent reservoir fluids (oil, gas, water) from entering the wellbore.
Performance Standard	Measurable statement, expressed in qualitative or quantitative terms, of the performance required of a system, equipment item, person or procedure (that may be part or all of a barrier), and that is relied upon as a basis for managing a hazard. The term includes aspects of functionality, reliability, availability and survivability.
Plug and Abandon	To prepare a well to be closed permanently, usually after either logs determine there is insufficient hydrocarbon potential to complete the well, or after production operations have drained the reservoir, or there are well integrity issues.
Pore Pressure	The pressure of subsurface formation fluids commonly expressed as the density of fluid required in the wellbore to balance that pore pressure. In reservoir zones which have sufficient permeability to allow flow, this is the pressure of the hydrocarbons or other fluids trying to enter the wellbore. Safe well design balances the reservoir pressure with drilling muds of sufficient density such that the mud hydrostatic pressure at the reservoir is sufficient to prevent inflow.
Prevention Barrier	A barrier located on the left-hand side of a bow tie diagram and lies between a threat and the top event. It must have the capability on its own to completely terminate a threat sequence. (Other possible name: **Proactive Barrier**).
Process Safety	A disciplined framework for managing the integrity of operating systems and processes handling hazardous substances by applying good design principles, engineering, and operating practices. It deals with the prevention and control of incidents that have the potential to release hazardous materials or energy. Such incidents can cause toxic effects, fire, or explosion and could ultimately result in serious injuries, property damage, lost production, and environmental impact.

Process Safety Management	A comprehensive set of policies, procedures, and practices designed to ensure that barriers to episodic incidents are in place, in use, and effective.
	The term is used generically in this document and is not restricted to the scope and rules of OSHA 29 CFR 1910.119 (frequently referred to as Process Safety Management or PSM). It is often aligned with the CCPS Risk Based Process Safety (RBPS) Guideline or the EI PSM Framework.
Risk	A measure of human injury, environmental damage, or economic loss in terms of both the incident likelihood and the magnitude of the loss or injury. A simplified version of this relationship expresses risk as the product of the likelihood and the consequences (i.e., Risk = Consequence x Likelihood) of an incident.
Risk Based Process Safety	The Center for Chemical Process Safety's process safety management system approach that uses risk-based strategies and implementation tactics that are commensurate with the risk-based need for process safety activities, availability of resources, and existing process safety culture to design, correct, and improve process safety management activities.
Risk Register	A regularly updated summary of potential major accident events over a facility life cycle, with an estimate of risk contribution and the barriers needed to achieve that level of risk. The risk register can be developed from facility PHA studies.
Root Cause	Management system failures, such as faulty design or inadequate training, that led to an unsafe act or condition resulting in an incident; underlying cause. If the root causes were removed, the particular incident would not have occurred.
Safeguard	Design features, equipment, procedures, etc., in place to decrease the probability or mitigate the severity of a cause-consequence scenario.

Safety Critical Element	Any part of an installation, plant or computer program whose failure will either cause or contribute to a major accident, or the purpose of which is to prevent or limit the effect of a major accident. Safety Critical Elements are typically barriers or parts of barriers. In the context of this book, safety includes harm to people, property and the environment. (Other possible names: **Safety and Environmental Critical Element, Safety Critical Equipment**).
Safety Case	A document produced by the operator of a facility which identifies the hazards and risks, describes how the risks are controlled, and describes the safety management system in place to ensure the controls are effectively and consistently applied.
Safety Culture	The safety culture of an organization is the product of individual and group values, attitudes, perceptions, and patterns of behavior that determine the commitment to, and the proficiency of, an organization's health, safety and environmental management. A Process Safety Culture extends the safety culture to address process safety issues equally to other safety issues.
Simultaneous Operations (SIMOPS)	Two or more different activities that are close enough to interfere with each other and transfer risk or performance implications. Examples include drilling and production operations or construction and production operations occurring at the same time.
Threat	A possible initiating event that can result in a loss of control or containment of a hazard (i.e., the top event). (Other possible names: **Cause, Initiating Event**).
Top Event	In bow tie risk analysis, a central event lying between a threat and a consequence corresponding to the moment when there is a loss of control or loss of containment of the hazard.
	The term derives from Fault Tree Analysis where the unwanted event lies at the 'top' of a fault tree that is then traced downward to more basic failures, using logic gates to determine its causes and likelihood.
Tripping Out	To remove the drill string from the wellbore.

Unconventional Reservoir	Unconventional oil or gas resources are much more difficult to extract than conventional. To be able to produce from these difficult reservoirs, specialized techniques and tools are used. For example, the extraction of shale oil, tight gas, and shale gas must include a hydraulic fracturing step in order to create cracks for the oil or gas to flow through. All of these methods are more costly than those used to produce fossil fuels from a traditional reservoir.
Underbalance	The amount of pressure (or force per unit area) exerted on a formation exposed in a wellbore below the internal fluid pressure of that formation. If sufficient porosity and permeability exist, formation fluids enter the wellbore. The drilling rate typically increases as an underbalanced condition is approached.
Workover	In many cases, workover implies the removal and replacement of the production tubing string after the well has been killed and a workover rig has been placed on location. Through-tubing workover operations, using coiled tubing, snubbing or slickline equipment, are routinely conducted to complete treatments or well service activities that avoid a full workover where the tubing is removed. This operation saves considerable time and expense. Other workover activities include casing repairs, clearing the well of sand, paraffin, hydrates or other substances that may form blockages.

ACKNOWLEDGMENTS

This Concept book was a joint effort between the Center for Chemical Process Safety (CCPS) and the Society of Petroleum Engineers (SPE). The American Institute of Chemical Engineers (AIChE), the CCPS, and the SPE express their gratitude to all the CCPS and SPE members of the subcommittee for their generous efforts and technical contributions.

The collective industrial experience and know-how of the subcommittee members makes this book especially valuable to all who strive to learn from incidents, take action to prevent their recurrence and improve process safety performance.

PROJECT AUTHORS

This manuscript was written by staff from DNV GL under the guidance of the CCPS project committee and Cheryl Grounds as the staff consultant. The committee wishes to thank DNV GL for their efforts and unique perspectives.

Dr. Robin Pitblado and Tatiana Norman

PROJECT TEAM MEMBERS:

Center for Chemical Process Safety

Michael Broadribb	BakerRisk
Jean Bruney	Chevron (Subcommittee Chair)
Christopher Buehler	Exponent
Rajender Dahiya	AIG
Eric Freiburger	Praxair (Subcommittee Co-Chair)
Cheryl Grounds	CCPS Staff Consultant
Anil Gokhale	CCPS
Kevin He	Shell
David Mohler	Delek US
Carolina Morales	Ecopetrol
Luis Rodrigo Garcia	Ecopetrol
Jason Nicholls	XTO Energy
Kevin Watson	Chevron
Tracy Whipple	BP

Society of Petroleum Engineers

John Gidley	Atwood Oceanics (retired)
Terrance Sookdeo	Baker Hughes
Brad Smolen	BP
Rebekah Stacha	Society of Petroleum Engineers
Charlie Williams	Center for Offshore Safety (Subcommittee Co-Chair)

Before publication, all CCPS books are subjected to a thorough peer review process. CCPS gratefully acknowledges the thoughtful comments and suggestions of the peer reviewers. Their work enhanced the accuracy and clarity of this concept book. Although the peer reviewers have provided many constructive comments and suggestions, they were not asked to endorse this book and were not shown the final manuscript before its release.

Peer Reviewers:

Juan Ignacio Alveriz Claramunt	YPF
Kiran Krishna	Shell
Prince Onuwaje	Nova Chemicals
Sara Saxena	BP
Valerie Wilson	AIG

ONLINE MATERIALS
ACCOMPANYING THIS BOOK

Although few of the figures in this book are shown in black and white and reduced in size to enhance readability, some of them are available in color and larger size in an online register.

To access this online material, go to:

www.aiche.org/ccps/publications/upstreamPS

Enter the password: PSIU2020

PREFACE

The project to produce this concept book was a joint collaboration between CCPS and the Society of Petroleum Engineers (SPE).

The Center for Chemical Process Safety (CCPS) was established in 1985 to protect people, property and the environment from major chemical incidents by bringing best practices and knowledge to industry, academia, the government and the public around the world. As part of this vision, CCPS has focused on developing and disseminating technical information through collective wisdom, tools, training and expertise from experts within the oil, gas, and petrochemical industry. The primary source of this information is a series of guideline and concept books to assist industry in implementing various elements of process safety and risk management. This concept book is part of this series.

SPE is a transnational technical and professional society serving members engaged in the exploration, development, production and mid-stream segments of the oil, gas, and related industries. It has a mission to collect, disseminate, and exchange technical knowledge concerning the exploration, development and production of oil and gas resources and related technologies for the public benefit.

As a not-for-profit organization, CCPS has published over 100 books, written by member company representatives who have donated their time, talents and knowledge. Industry experts, and contractors that prepare the books, typically provide their services at a discount in exchange for the recognition received for their contributions in preparing these books for publication.

1

An Introduction to Process Safety for Upstream

1.1 BACKGROUND

This concept book on the topic of process safety for the upstream oil and gas industry is aimed at both new and experienced members of the upstream industry and for those thinking of transferring to the industry from a downstream or midstream position or another industry. CCPS concept books differ from CCPS guideline books. Concept books are written at an introductory level as compared to guideline books which aim to explain in detail methods that are well developed and accepted by most industry companies. Readers should be able to implement the topic of a guideline book for their specific circumstances. Concept books are shorter and set out sound methods in outline but may not yet be fully embedded by the target audience. They lay the foundation for potential future guideline books addressing individual topics in greater detail.

The term process safety is taken from the CCPS Glossary of Terms, and this and other relevant terms are in the Glossary of this book. It is defined as:

"A disciplined framework for managing the integrity of operating systems and processes handling hazardous substances by applying good design principles, engineering, and operating practices. It deals with the prevention and control of incidents that have the potential to release hazardous materials or energy. Such incidents can cause toxic effects, fire, or explosion and could ultimately result in serious injuries, property damage, lost production, and environmental impact."

Process safety focuses on loss of containment events that can cause serious harm to people, the environment and to the asset. These are often denoted by the acronym LOPC – Loss of Primary Containment events. It does not explicitly address other causes of major consequence events which might be due to harsh weather, failure of marine systems, transportation, etc., unless these subsequently cause a loss of containment incident. Nor does the definition explicitly mention the consequence of loss of reputation, but most companies consider this as well when ranking risks. The holistic term for all large potential incidents from whatever cause is often "major accident events".

Fundamental to the topics presented in this book is process safety management using the framework of CCPS *Guidelines for Risk Based Process Safety* (RBPS) (CCPS, 2007a). This sets out a structure of four pillars and twenty management system elements and is described more fully in Chapter 3. A definition for process safety management systems, also from the CCPS Glossary, is:

"Comprehensive sets of policies, procedures, and practices designed to ensure that barriers to episodic incidents are in place, in use, and effective."

It is important to differentiate the scope of process safety from personal or occupational safety. Process safety focuses on infrequent or rare events which have severe consequences, compared to personal or occupational safety, which aims to prevent more frequent events with lesser consequences usually limited to a single individual (Figure 1-1). This book focuses on the former, not the latter. This differentiation is necessary as the required safety solutions and management systems are different. More frequent occupational safety events are typically easier to anticipate and manage than rare process safety ones, which may not be experienced in the lifetime of a single facility. Many historical examples exist where facilities with excellent occupational safety records experienced serious process safety events. Subsequent investigation showed large gaps in managing the potential for major incidents. A well-known example of this is the blowout on the Deepwater Horizon (National Commission, 2011).

1.2   APPLICABILITY OF PROCESS SAFETY TO UPSTREAM

Some people think that process safety applies only to chemical processes in the downstream sector, because this is where the terminology originated. Process safety

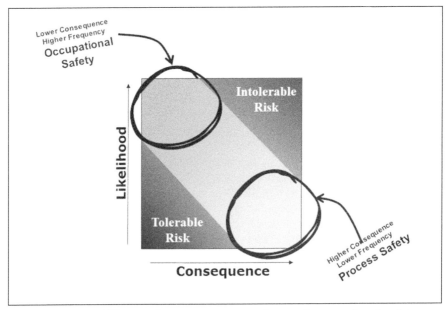

Figure 1-1. Difference between process safety and occupational safety

is the holistic approach to major incidents, their prevention, preparedness for, mitigation of, response to, and restoration from serious releases of hazardous substances or energy from facilities. Process safety applies equally to upstream and downstream.

A loss of containment event can occur in most parts of the upstream production system and operations life cycle. Containment can be lost from a well during drilling, completion, production, workover, intervention or abandonment. It can also be lost from production facilities, whether onshore or offshore, and from any associated storage or transportation. Upstream also covers gas plants and mid-stream plants (e.g., LNG). LNG facilities are very similar to many downstream facilities.

The term used to describe process safety is called different names in different industries and this may be confusing. In aviation, space and nuclear industries, the term used is system safety. Historically, system safety was used to show how complex systems interact and potentially fail. Offshore, sometimes the term used might be major incident safety, operational integrity, loss prevention, well control, technical safety, operational risk management, or safety and environmental management. The proliferation of terms causes confusion and this book will use the term "process safety". For reasons of space, this book does not address major incidents due to weather, ship collision or structural failure

In this book, multiple examples of incidents affecting the upstream industry are presented and these show the applicability of process safety to upstream. For example, a well-known onshore incident is the Pryor Trust incident (summarized in Chapter 4) and for offshore the Piper Alpha incident (summarized in Chapters 3 and 6). Multiple aspects of the RBPS framework are applicable to both incidents and implementation might have contributed to their prevention or mitigation.

1.3 INTENDED AUDIENCE

This book has as its primary intended audience the following potential readers.

- Process safety professionals, project engineers and design engineers, health, safety and environment (HSE) professionals, and other personnel in the upstream industries with process safety responsibilities (such as but not limited to, asset leadership team, project engineers and their managers, design engineers, and safety representatives / professionals)
- Young professionals or newcomers just entering the upstream industry or those transferring from the downstream or midstream industry, or another industry
- Experienced personnel seeking a credible external reference
- Process safety consultants just entering the upstream industry
- Personnel in rig companies, service companies, and supply companies

- Personnel in engineering and construction companies
- Personnel in supply chain and equipment suppliers
- Regulators (both offshore and onshore)
- Informed stakeholders such as potentially affected public

The book is not targeted at operators or maintenance technicians on an upstream installation. However, they should benefit from the application of process safety concepts via the audience listed above, even though they might not read this text themselves. The International Association of Oil & Gas Producers (IOGP) and Center for Offshore Safety (COS – an entity within API) have several publications addressing process safety, including some targeting process operators and maintenance technicians.

Being on the above list does not imply that those groups are not familiar with process safety concepts, but this book may help using the concepts in a standard manner, with consistent terminology, and with up-to-date practices.

1.4   WHY THE READER SHOULD BE INTERESTED

There are good business reasons for applying process safety concepts to upstream operations. The *Business Case for Process Safety* CCPS (2018) sets out many important benefits beyond improved process safety, with additional economic returns related to the following.

- Corporate responsibility
- Risk reduction
- Business flexibility
- Sustained value

The business case thinking is consistent with a philosophy that establishes safety as a "value" rather than a "priority". A value is a core belief that does not change, whereas a priority balances with other priorities and may be emphasized or deemphasized depending on circumstances.

A feature of process safety is that it focuses on infrequent or rare events that have serious potential consequences. It is quite possible that a facility could operate for many years without applying the concepts of process safety and not experience a major incident simply due to the low probability of process safety events. Based on past experience, a major incident may occur at some point over a population of multiple facilities (e.g., a large company or a geographic region) with serious consequences to people, the environment, the asset, and to company reputation. So, process safety is typically not something justified by short term economic gain; however, it is the right thing to do and benefits long term sustainability.

The concepts of process safety management described in this book also have a benefit on facility efficiency – improved reliability, increased up-time, fewer incidents, and better resilience (i.e., the ability to recover from significant deviations). These benefits accrue from formally assessing potential process safety deviations and introducing barriers to prevent or mitigate the consequences. These barriers also reduce near misses, less serious incidents, and operational challenges, and hence improve uptime.

Incidents can have serious economic costs. Often the facility must be shut down for an extended period as investigations are carried out, repairs are completed, and findings may identify process safety changes to be introduced before the facility is restarted. Table 1-1 provides some examples of major process safety incidents that have occurred in the upstream industry.

A good overview of major incidents and lessons learned is provided in *Incidents that Define Process Safety* (CCPS, 2008b) and *More Incidents that Define Process Safety* (CCPS, 2019). While most examples in these refer to downstream, there are some for upstream as well. Additional sources of major incident information include Marsh (2020) and with more detail in Lees (2012).

Table 1-1. Selected incidents
(from Marsh, 2020 and National Commission, 2011)

Incident	Year	Short Description of Incident
Enchova, Brazil	1988	A blowout was ignited, and a resultant fire persisted for 21 days. Most of the facility was destroyed.
Piper Alpha, North Sea	1988	Poor permit management led to a condensate leak and initial explosion, followed by pool fires and a major jet fire. The escalation led ultimately to 167 fatalities and total loss of facility.
Longford, Australia	1998	A brittle fracture of a heat exchanger led to 2 immediate fatalities and a long-lasting fire at a critical location. The loss of gas supply to the State of Victoria led to thousands of temporary layoffs.
Mumbai High, India	2005	A support vessel collision caused a leak and fire, 22 fatalities and total loss of facility, but adjacent bridge linked facilities survived.
Montara, Timor Sea, Australia	2009	A blowout event persisted for 2 months until a relief well was completed. The rig ignited just as heavy mud was being introduced to kill the event. The facility was a total loss.
Deepwater Horizon, USA	2010	Blowout event which caused 11 fatalities and led to the largest oil spill in US history. The well was capped successfully after 3 months and no further loss of containment occurred.

This selected list of incidents with loss of containment shows how serious major process safety events can be to people, the environment, and to business. Marsh (2020) lists many serious upstream losses in its 100 largest losses review. This source lists major losses due to process leaks (e.g., Piper Alpha), blowouts (e.g., Deepwater Horizon), harsh weather (e.g., Ocean Ranger), failure of marine systems (e.g., Kolskaya towing incident), structural failures (e.g., Alexander Kielland leg collapse and sinking) and transportation (e.g., helicopter incidents). Details on all of these are available from the relevant regulator or by a literature search. While this book focuses on process safety, readers can learn from all incidents in efforts to improve overall upstream safety.

Although this book does not delve into specific regulations, certain regions require process safety and other major hazards to be addressed as part of the permitting process. Examples include the US SEMS and OHSA 1910 rules, Europe (EU) Offshore Oil & Gas Safety Directive, Norway PSA requirements, Australia NOPSEMA rules, and Abu Dhabi ADNOC requirements. The implementation requirements are different, but all require major hazards to be identified, assessed for risk, and managed with an effective safety and environmental management system. A short summary of international regulations is provided in Section 2.8.

This book is needed for several reasons.

1. Major incidents in the upstream industry such as Piper Alpha (IChemE, 2018) and Deepwater Horizon (National Commission, 2011) show that robust process safety management is beneficial not only to reduce events but also to demonstrate to the public that the industry is managing its risks effectively. This latter aspect is important for the community and regulators to have confidence in the industry and thus allow continued or new operations. This book is intended to help improve process safety performance, thus supporting the industry as a whole. A similar argument applies to the downstream industry as well.

2. Newcomers to the industry can benefit from a text specifically explaining process safety in the context of the upstream industry. SPE has a substantial library of books in its textbook series. These focus on technical aspects of well design and upstream operations, rather than process safety. This book fills a gap in SPE literature.

3. The upstream industry can leverage learnings from both its own incidents and those from downstream related to hard-won lessons of major incidents such as Flixborough and Bhopal (both described in Lees, 2012). These lessons have been codified by CCPS in a series of over 100 Guideline texts addressing process safety, including *Risk Based Process Safety* (CCPS, 2007a). CCPS was created as the US downstream industry response to the Bhopal disaster in 1984. It has had its objective to put into the public domain the best practices for process safety. Most of these are equally applicable to upstream as well, although the technical terms and examples may differ. This book is an access point to many of the other CCPS texts.

4. This book can help upstream personnel improve their understanding and communication of the concepts of process safety management.

1.5 SCOPE OF THIS BOOK

The upstream oil and gas industry is diverse. This concept book provides an overview of process safety as it applies in the upstream industry. Hopefully, this book will spur the interest in developing subsequent, more detailed, guideline texts.

After this Introduction chapter, this book provides an overview of upstream operations in Chapter 2, including an introduction to safety barriers, and a short review of international regulations. This is followed by a summary of Risk Based Process Safety (RBPS) in Chapter 3, along with short descriptions for each element. The book then covers the application of the various RBPS concepts throughout upstream operations: well construction (both onshore and offshore) in Chapter 4, onshore production in Chapter 5, offshore production in Chapter 6, engineering design, construction, and installation in Chapter 7, and future topics and research needs in Chapter 8.

As noted earlier, the focus is on process safety (i.e., loss of containment events), so other major incident hazards (adverse weather, marine events, structural failure, transportation incidents) which would require extensive discussion, are not covered in this concept level book. However, these events can be initiating events for loss of containment – e.g., the Mumbai High event in Table 1-1. The methods described in this book can be applied to these other major incident hazards as well. Similarly, occupational safety is not addressed other than toxicity or fire and explosion that can affect many people at once.

Liquefied Natural Gas (LNG) can be thought of as upstream or midstream. In this book, it is covered briefly where the liquefaction occurs offshore in floating facilities (FLNG units), but not onshore in full scale liquefaction plants as these are very similar to downstream facilities and thus are already covered in the existing CCPS library.

A figure showing the topics which are in scope and those that are not in scope is shown in Figure 1-2. The column topics are addressed in Chapters 4, 5, 6, and 7 as indicated.

1.6 UPSTREAM SAFETY PERFORMANCE

Upstream incidents are tabulated by several organizations. Offshore, individual regulators collect their own safety data (e.g., BSEE, UK HSE, PSA, etc.). They do their own reporting, but also share this information to the International Regulators Forum (https://irfoffshoresafety.com) allowing for easier comparison between regions using standardized categories. In the US, upstream onshore activity is less

Lifecycle Stage	Well Construction (Chapter 4)	Onshore Production (Chapter 5)	Offshore Production (Chapter 6)	Engineering Design (Chapter 7)
In-scope	Onshore drilling Offshore drilling Onshore completions Offshore completions Workovers Abandonment Well control Surface operations Subsea equipment	Wellheads Gathering lines Gas plants Storage tanks Export operations Traditional / Fracking SIMOPS Decommissioning	Shallow or deepwater Surface operations Subsea processing Export options FLNG installations SIMOPS Decommissioning	Design stages FEL 1-2-3 Inherent safety Process safety Well integrity
Ex-scope	Seismic surveys Staff transportation	Transport to customer - pipeline, truck, train Staff transportation Occupational safety LNG Plants	Transport to customer - pipeline, ship Staff transportation - Helicopter, boat Occupational safety Marine incidents	Construction safety

Figure 1-2. Scope of *Process Safety in Upstream Oil and Gas*

formalized for process safety than offshore. Large scale onshore upstream operations are covered by PSM and RMP, but smaller developments are not covered by these federal process safety regulations (CSB, 2018) nor are drilling activities of any size. The regulatory focus for smaller onshore operations is occupational safety and environment with state and local regulations predominating. Process safety is driven by following relevant API standards. Upstream onshore operations have a larger number of process safety incidents than offshore (as is shown in Section 1.6.3) but these usually have fewer impacts to people, and this may be a factor in the degree of regulation.

1.6.1   Analysis of US Offshore Safety Data

Halim et al (2018) analyzed Bureau of Safety and Environmental Enforcement (BSEE) incident reports. Offshore operators within the US Outer Continental Shelf are required to report specific incidents to BSEE. Over the period 1995-2017, a total of 1,617 incidents were investigated. The authors further analyzed 137 fire and explosion incidents over the period 2004-2016 where there was sufficient detail to establish causation. They identified nine of the most common causes, of which equipment failure and human error dominate.

COS (Center for Offshore Safety) also provides incident data reported by its membership. Smolen (2019) summarizing this data shows while process safety performance has improved over several years in some categories, it may be plateauing. COS also collects Tier 1 and Tier 2 process safety incident data.

1.6.2   International Incident Data from IOGP

Incident trend data is available from IOGP (International Association of Oil & Gas Producers) covering both onshore and offshore incidents. IOGP is a consortium of companies which operates in 80 countries and collectively produces about 40% of global oil and gas. They have produced an annual report on safety performance trends since 1985. IOGP (2019a) in a summary figure (Figure 1-3) shows trends on two process safety categories: Tier 1 (larger events) and Tier 2 (smaller but still serious events). Tier 2 events are declining, but Tier 1 appears flat.

Fatal accident trends have decreased by about 70% in the most recent 10-year period (2008-2017), with Fatal Accident Rate (FAR) decreasing from 3 to 1 fatality per 100 million hours worked. Over a longer period from 1985 to 2017 the FAR has reduced from 18 to 1 fatality per 100 million hours.

Most fatality events are single fatalities – for example, 28 of 30 fatal incidents reported in 2017 involved only one person and similar trends exists for other years. In terms of activity category causing fatalities, drilling was the highest, with transportation on land and maintenance being the next highest categories.

IOGP issued a guideline on reporting process safety performance indicators (IOGP 456, 2018a). This guideline follows closely the format of performance indicators developed for downstream in API 754. Both define a set of leading and lagging indicators in four tiers. Tiers 1 and 2 are incidents meeting specific consequence thresholds (harm or damage) to or releases of a specified substance and size. Tiers 3 and 4 are omissions of safety activities or demands on safety systems and management system deficiencies. Because the two guidelines are aligned providing worldwide consistency, some companies are starting to report on the number of Tier 1 and 2 events in their annual reports covering all their operations. As the company annual reports usually combine these process safety statistics for the entire integrated business, it is not easy to establish trends for the upstream sector alone from this source.

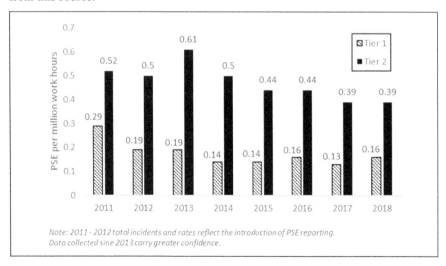

Figure 1-3. Tier 1 and Tier 2 process safety events
Source: IOGP (2019a)

IOGP provides a comparison of onshore and offshore process safety events in its 2019 report. It is shown that onshore activities (both drilling and production and normalized for the number of manhours) have significantly higher Tier 1 and Tier 2 rates than offshore activities. Throughout the period 2011 to 2018, onshore was 2.5 times offshore for Tier 1 events and 1.5 times for Tier 2 events and for the US this ratio was even higher. However, in the next section summarizing the Marsh 100 Largest Losses analysis, no onshore upstream incident appears in the list of 100 events, but 23 offshore upstream events do.

1.6.3 Marsh 100 Largest Losses

Marsh (2020) publishes incident data periodically on the 100 largest losses – based on the financial cost. Since the list is always 100 incidents, with newer larger losses updating smaller inflation corrected older ones, it is not a good source for statistical analysis, but it does provide a breakdown by five industry types: refineries, petrochemicals, gas processing, terminals and distribution, and upstream.

An examination of the upstream category shows that 23 offshore incidents appear in the list but no onshore incidents. This is true even though IOGP shows the onshore incident rate is significantly higher than the offshore rate. Marsh does not provide an explanation for the difference, but it may be speculated that offshore facilities are often larger, more densely packed and stacked vertically, and with personnel accommodation onboard which can lead to larger losses. Also, the financial investment in offshore deepwater facilities is often much greater than onshore, so it is unsurprising that Marsh's largest losses will be offshore.

Marsh notes that most of the largest losses in last two years are in the downstream sector, from refineries and petrochemical assets, particularly those built in the 1960s or earlier. Refineries account for 50%, petrochemicals for 25% and gas processing and terminals for the balance of the new losses. Marsh highlights contributing factors to the downstream spike include a reduction in global engineering standards (specifically not updating older equipment to meet newer standards), less stringent regulation, the higher utilization of certain facilities such as refineries, and aging infrastructure. Upstream is not immune to these factors.

In a prior report, issued in 2018, Marsh identified a need for improvements to systems of work (e.g., permits to work, shift handover communications, and management of change) and to inspection (e.g., staffing levels, competency, and data analysis). Across all areas in their report Marsh identified mechanical integrity as a key area of causation. Other contributors included inadequate hazard identification, inadequate risk assessment of critical tasks, reliance on remotely operated valves for safe isolation, and failure to identify safety critical devices.

1.7   SUMMARY

This concept book aims to provide an overview of process safety for the upstream oil and gas industry. It suggests the intended audience and the reasons why the book is needed. The scope of this book includes well construction, onshore and offshore production operations including gas processing, some aspects of LNG, engineering design, construction and installation and future needs. However, occupational safety and major hazards not related to process safety are excluded from this book as they are well covered in other industry guidance.

Several serious upstream process safety events have been identified and more are used throughout the rest of the chapters to reinforce application of RBPS concepts. While the downstream industry has also suffered many serious process safety events, on average upstream (offshore) events have higher impacts to people due to the numbers of personnel located in close proximity to major hazards and with potentially limited choices of egress in case of the need to evacuate from an incident.

A collection of upstream incident statistics was presented. These reinforce the importance of process safety and risk reduction efforts in the upstream oil and gas industries. While occupational safety performance is improving, progress on process safety may be plateauing and warrants increased focus.

2

The Upstream Industry

2.1   UPSTREAM INDUSTRY

2.1.1 Life Cycle Stages

This chapter provides an overview of the upstream industry and follows the upstream life cycle. Greater details on well construction, well work, and onshore and offshore production are provided in Chapters 4, 5 and 6, respectively. Engineering design, construction, and installation are covered in Chapter 7. The main life cycle stages are shown in Figure 2-1 and are described as follows.

- Exploration and Well Construction, including discovery and appraisal wells
- Engineering design, construction and installation
- Production phase (covering first oil, build-up, plateau and decline)
- Well workovers and interventions to maintain well integrity and boost production during decline
- Decommissioning / abandonment

The complete life cycle is more complex than the figure shows as additional wells may be constructed and well stimulation activities or other enhanced oil recovery methods may be implemented to maintain production levels.

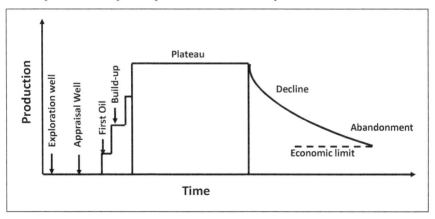

Figure 2-1. Typical upstream life cycle

The upstream industry exploration and production systems are diverse and complex, with many different designs to accommodate well conditions, variation in well fluid characteristics, processing conditions and water depth, if offshore. The scope of the upstream industry that falls within the scope of this book is mentioned previously in Figure 1-2 in Chapter 1.

Wells both onshore and offshore produce oil, gas and/or condensate along with associated non-hydrocarbon gases, water and sand. The composition affects the well design, the surface operations, and the potential for process safety events.

Conventional reservoirs require a source rock containing hydrocarbons which migrate through porous rock or through fracture planes to a trapped area with an impervious seal rock where they accumulate. Absence of an impervious seal rock means any oil and gas dissipate and no reservoir forms. Good descriptions of important reservoir characteristics are provided in Franchi and Christiansen (2016).

Gas and condensate are light and often free flowing to the surface. Oil is heavier and may be free flowing or require artificial lift for its recovery. Due to high pressures in the well, gas and condensate may remain dissolved in the oil phase until the pressure drops near the surface. Crude oils are typically characterized by API gravity, which relates to specific gravity (SG).

2.1.2   Types of Upstream Facilities

Upstream operations begin with exploration wells, usually after seismic surveys and interpretation of prior drilling at nearby wells, if available. If exploration drilling identifies a commercial prospect, then the well may be converted to a production well. Alternatively, it may be permanently abandoned, and a new production well drilled with specific casing diameters and lengths to match the production profile of the better understood reservoir characteristics and anticipated flowrates.

There are a wide variety of drilling and production facilities located onshore and offshore. ABB (2013) provides some useful examples.

Large onshore fields may have many wellheads with outlet product flowlines that are then manifolded into gathering lines which merge with other lines and then flow into a large central product treatment facility (details on treatment are provided later in this section). Onshore facilities in warmer climates may be in the open, but if located in cold climates (e.g., Alaska), then these are typically housed in large buildings to provide weather protection for personnel and equipment.

Smaller onshore fields are often of limited scale and are located on pads containing multiple wells. This is particularly true with wells that require fracturing. These may be connected by gathering lines to larger sites. There may be basic treatment facilities with further processing to occur at other sites. There may also be local storage tanks for liquid products awaiting export by truck, train or pipeline.

Offshore facilities vary greatly in design due to water depth and the harshness of the climate. The *Petroleum Engineering Handbook* (2007) and ABB (2013) both provide more detailed introductions to upstream operations than is possible in this chapter. Also, the SPE Technical Library provides multiple books on many aspects of well design and operation, and offshore production.

Table 2-1 provides an indication of the range of drilling rigs and production facilities onshore and offshore and provides links to photographs of these facility types. This is not an exhaustive list, but it covers most facility types.

Table 2-1. Types of upstream rigs and facilities

Location	Characteristics	Description
Onshore	Large integrated treatment facility (see Figure 2-2)	These facilities separate oil and gas remove unwanted materials (e.g., produced water), stabilize the crude oil, and export the oil and gas, usually by pipelines. If no gas export pipelines are available, the gas is compressed and reinjected.
Onshore	Smaller onshore drilling rig (see Figure 2-3)	Drilling rig for vertical or directional drilling. Conventional reservoirs may be free flowing. Some wells may need additional treatments such as fracking or enhanced oil recovery to increase production.
Offshore – shallow water	Jack-up drilling rig (see Figure 2-3)	These are usually 3-legged floating hulls with legs that can be jacked-up to allow for free movement and jacked-down to provide a stable platform for well construction. These are often used in 300-400 ft of water depth due to practical limits on leg height.
Offshore – shallow water	Jacket / Platform (see Figure 2-4 and Figure 2-9)	These jackets and platforms support the processing facilities and any necessary accommodation. Accommodation(s) are sometimes located on other structures linked by a bridge. Pipelines export oil and gas to shore.
Offshore - deepwater	Drillship or semi-submersible rig	These are floating drilling rigs (either ship-shaped or semi-submersible) which use GPS-based dynamic positioning to hold position or an anchoring system while conducting well operations.

Location	Characteristics	Description
Offshore - deepwater	Spar or Tension Leg Platform (see Figure 2-5 and Figure 2-9 for TLP and Figure 2-8 for SPAR)	These are buoyant processing and accommodation structures, held in place by mooring lines (for spars) or vertical tendons (for TLPs).
Offshore - deepwater	FPSO – Floating Production Storage and Offloading (see Figure 2-6 and Figure 2-8)	These are ship-shaped vessels, often repurposed oil tankers, with processing facilities mounted on the upper decks and storage in tanks within the hull. Usually the vessel weathervanes freely around a turret which is held in place by mooring lines and anchors. The turret contains the risers and other links to the wellhead located on the seabed. Some designs do not use a turret, and instead moor the vessel in a fixed orientation.
Offshore - deepwater	FPU – Floating Production Unit (see Figure 2-7 and Figure 2-8)	FPUs are also known as semi-submersible designs. These are large facilities mounted on vertical columns extending down to horizontal pontoons. Processing facilities and accommodation are located on multiple decks. Export is direct to pipeline, usually with no storage on the FPU.

Figure 2-2. Example onshore treatment facility, Alaska

Figure 2-3. Example onshore drilling rig (left) and offshore jackup rig (right)

Figure 2-4. Example fixed platform with linked bridge (North Sea)

Figure 2-5. Example Tension Leg Platform (TLP) (Gulf of Mexico)

Figure 2-6. Example FPSO

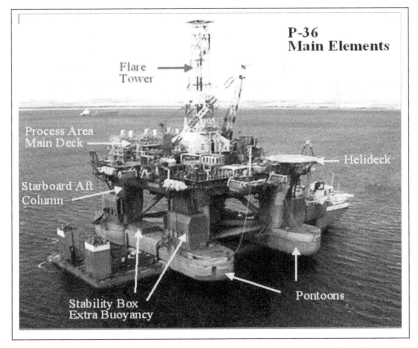

Figure 2-7. Example FPU (on dry tow showing parts normally submerged)

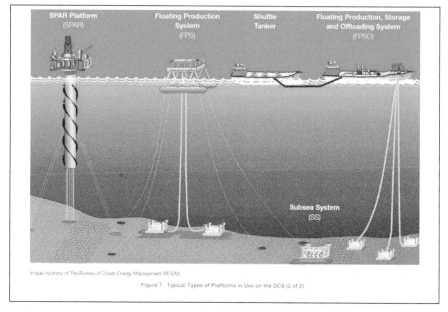

**Figure 2-8. Typical types of production installations in use on the deepwater outer continental shelf (OCS)
Courtesy of BOEM**

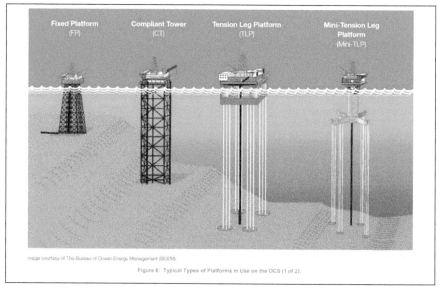

**Figure 2-9. Typical types of production installations in use on the outer continental shelf (OCS)
Courtesy of BOEM**

2.2   EXPLORATION PHASE

2.2.1   Onshore

The exploration phase refers to drilling carried out where the primary information available is from seismic surveys or from nearby wells. The terms conventional and unconventional are sometimes used to describe reservoirs. A conventional reservoir is one where buoyancy keeps oil and gas sealed beneath a caprock and these flow easily into a wellbore for production. Unconventional reservoirs differ in that fluid trapping mechanisms and other formation properties differ to conventional reservoirs and require different means to produce. Examples include coalbed methane, shale gas, tight gas, and tar sands. The main process safety concerns with well construction during exploration are a loss of well control leading to a blowout event, a leak into subsurface water layers, or a leak from surface operations. Onshore wells can have a direct safety and environmental impact to the public, depending on their location. Offshore wells are primarily a safety issue for their workforce, but there can be public safety and environmental impacts as well.

Exploration wells have the objective to determine the extent of the hydrocarbon deposit and its composition (oil or gas), temperature, pressure and potential flow rate and other well characteristics. Information regarding local geology and hydrocarbon pressure can also be gathered to improve subsequent drilling and well control decisions. Multiple exploration wells or appraisal may be needed to establish the size of a reservoir and whether it is commercially viable.

Key factors in well control are the local geology and pore pressure of each strata in the wellbore. Reservoir properties and how they affect casing requirements and mud weight are described in more detail in Chapter 4.

Onshore wells can be economically viable with relatively smaller hydrocarbon reserves than offshore wells. This is because the infrastructure required is usually much less costly than offshore (i.e., a well pad is required but no jackets or floating structures) and the workforce may be able to drive directly to the site without the need for helicopters, marine transport, or offshore accommodations. However, some onshore wells are in remote areas (arctic, desert, or jungle locations) and might require helicopter transport. Accommodation, if needed, is usually adequately separated from the well construction activity.

While some wells may still be associated with conventional large deposits which can be exploited using traditional vertical or directional drilling, the fastest growing area onshore is for shale fields. Shale wells employ horizontal drilling once the hydrocarbon formation is reached. High-pressure hydraulic fracturing or fracking of the shale layer is used to open higher flow potential channels and release the trapped oil and gas.

Another type of unconventional reserve is oil sands as may be found in Venezuela or Alberta, Canada. Although these deposits are very large, the technology to recover the oil is costly as it may involve mining or heat treating the field to recover the oil. Transport of heavy oils for further processing may require diluent (e.g., kerosene) to make the oil flow more freely in the pipeline and a second pipe to return recovered diluent back to the well site to repeat the cycle.

Process Safety Issues

Key process safety issues associated with onshore exploration include blowouts (including shallow gas blowouts), hydrocarbon intrusion into freshwater aquifers, and hydrocarbon loss of containment from surface facilities. Further details are provided in Chapter 4 for drilling and Chapter 5 for onshore production. Blowouts are prevented by active well management and early detection of well kick events that signal a potential influx of hydrocarbons into the wellbore. Primary process safety controls are (1) the mud column and (2) the blowout preventer (BOP) and the well pressure containment system. The mud column uses high-density mud to hydrostatically prevent formation fluids from entering the wellbore. The BOP consists of several valves/rams designed to close off the wellbore to control potential blowouts. The BOP is placed on the surface for onshore drilling, and at the surface or on the seabed for offshore. Drilling is usually carried out by specialist contractors and there is a need for good communication of process safety between the owner/operator and the drilling contractor, as is noted in IADC guidance.

2.2.2 Offshore

Drilling offshore is similar to onshore, but the drilling rig is different. In shallow water up to 300-400 ft (90-120 m) drilling is carried out from a fixed platform or using a jack-up rig. A fixed platform cannot be moved once installed, whereas a jack-up is mobile. The jack-up legs have large spud-cans (sometimes with mats) on the bottom of each leg. Spud-cans use the weight of the rig and ballast water to penetrate the mud on the sea floor to provide a stable platform for the rig while conducting well operations. Jack-ups can be self-propelled, but more often they require tugs to move them between locations.

In deeper water, well operations are accomplished by floating drill ships which are ship-shaped, semi-submersibles (i.e., derrick and decks supported by columns onto pontoons providing most of the buoyancy), or platform-based rigs deployed on production facilities. Drill ships are generally self-propelled whereas most semi-submersibles are towed between drilling locations. Position is held during well operations either by an array of anchors or more commonly by dynamic positioning (multiple GPS-controlled thrusters). BSEE normally considers deepwater well operations to be in 1000+ ft (305+ m) or greater of water depth.

Wells drilled in deep water generally have much greater total lengths than normally seen onshore or in shallow water because of the distance to the seabed. The cost of deepwater well operations are higher than onshore or shallow water operations, mainly due to cost of the rig, logistical support and time.

Process Safety Issues

The mud column and the BOP and well pressure containment system are the primary controls for loss of well control. These are operated in combination with active well monitoring and effective well control management. BOPs may be located on the drill ship at the surface or in deep water, just above the seabed. The reliability of these devices is critical, especially if located on the seabed as they are hard to access and maintain and may be positioned there for several months. The mud column and BOP are discussed in more detail in Chapter 4.

2.2.3 Completion

Once the well is drilled into the reservoir, several activities are required to complete the well. Final casing and cementing provide well integrity and production tubing is installed along with other downhole tools. Typically, the casing is perforated at designated locations in the hydrocarbon layer using explosive charges to provide for flow into the well.

Process Safety Issues

The main process safety issues associated with completion are well control and failure to achieve well integrity with the final casing and cementing (see further discussion in Chapter 4).

2.3   ENGINEERING DESIGN, CONSTRUCTION AND INSTALLATION

2.3.1 Engineering Design

Once the exploration wells and any further appraisal wells are drilled and the characteristics of the reservoir are determined, the next stage is the design, construction and installation of the production system. Reservoir engineers, petrophysical engineers, and petroleum geologists determine the recoverable hydrocarbon reserves. Working with drilling engineers and production engineers, they determine how best to extract these. Civil and facility engineers take into account the location, and if offshore, the water depth and weather conditions. The reservoir characteristics determine the nature of the production systems required and the design of the production wells – which may or may not be any of the exploration wells. Onshore and offshore designs are quite different, reflecting the greater space available onshore versus the limited footprint possible offshore.

Key process safety design decisions are taken during the design stage, and once taken it may be difficult to significantly improve process safety design during operations. Inherent safety is addressed at this stage. For process safety design, key aspects are to limit both the inventory of hazardous materials and the number of people required in proximity to the hazards on or at the facility. This is even more critical offshore where available space is limited. Additionally, operability, ease of maintenance, and instrumentation, monitoring, and shutdown systems, and personnel facility access are process safety design topics.

Often several design options are possible onshore and offshore, and the final investment decision is only taken after a stepwise analysis. CCPS (2019b) notes the overall design process is termed Front End Loading (FEL) and is usually broken into 3 stages: FEL-1 Appraise, FEL-2 Select, and FEL-3 Define. The means to address process safety issues during engineering design is covered in Chapter 7.

Requirements for process safety studies are affected by the jurisdiction, for example the requirements are different in Norway, Australia and the US (see Section 2.8). The UK and EU are currently the same (in 2020) but in future may differ as the UK is no longer part of the EU. CCPS and SPE endorse a risk-based approach consistent with CCPS RBPS, regardless of specific jurisdiction minimum requirements.

Process Safety Issues

CCPS (2019b) summarizes FEL-1 process safety activities as including a preliminary Hazard Identification (HAZID), a preliminary Inherently Safer Design (ISD) review, a concept risk analysis, development of an HSE and process safety plan, a risk register, and follow-up action tracking. FEL-2 includes refining the FEL-1 studies and additionally recommends further process safety studies, such as hazard identification and risk analysis, fire and explosion analysis, blowdown and depressurization study, and fire and gas detection study. A final design decision is normally taken at the end of FEL-2. These studies aid the selection of the best design option. FEL-3 (also known as FEED – Front End Engineering Design) refines these studies further based on the final selected design. A PHA is performed at this stage when adequate information is available such as P&IDs. The regularly updated risk register ensures that no identified risk is overlooked and that the design team addresses all recommendations. See Chapter 7 for further details on these process safety studies.

2.3.2   Construction and Installation of Production Facilities

Onshore

Onshore production facilities may be in the open, or if in harsh climates, inside large modules. Construction is not very different to downstream facilities; however, issues may arise if the production facility is already in operation for other wells (see later discussion of SIMOPS in process safety issues). Often large onshore fields are in remote locations and this can pose difficult logistical issues and worker occupational safety issues. Roads may need to be constructed for access and this can cause environmental issues. In the arctic, in order to avoid damage to permafrost soils, heavy equipment may be restricted to winter movements only when the ground is frozen.

Offshore

Offshore production facilities are usually constructed at specialized yards and later integrated with processing modules which may be assembled elsewhere. This applies whether the facility is a steel jacket or a floating deepwater facility. An FPSO may be constructed in two different yards – a yard for the tanker hull conversion and a yard to integrate all the topside processing facilities. The customer defines the requirements and specialist design companies or yard in-house designers design the facility in consultation with the customer. Offshore facilities, if they are capable of being self-propelled (e.g., an FPSO), are subject to marine classification and must follow design requirements specified by one of several Class Societies (e.g., ABS, DNV GL, Lloyd's Register, etc.). Even where not required, some companies may choose to implement Class rules for fixed structures as an additional check on the design. Also, non-propelled floating structures may have other marine specific requirements as they do active ballasting and other marine activities.

Fixed platforms may be constructed onshore in the yard and towed to their final location, or they may be constructed in pieces and assembled at their final location. Floating deepwater facilities are completed in the yard and either are sailed or towed to their final location.

Process Safety Issues

Process safety issues are limited during construction and installation whether onshore or offshore, although issues relating to occupational or construction safety are different. Significant process safety issues will exist if there are simultaneous operations (SIMOPS) with the interaction between existing operations at a site or on an offshore facility and construction activities that could lead to hazards not normally present when only one activity is underway. A range of process safety issues are discussed more fully in Chapter 5, 6 and 7.

2.4   PRODUCTION PHASE

Production activities onshore and offshore are broadly similar, but with some important differences. Production rates onshore can be very large (e.g., Alaska or Saudi Arabia), but also relatively small (e.g., shale field developments or mostly depleted reservoirs). Onshore plants usually have local storage for produced liquids whereas offshore liquids are often immediately exported by pipeline. Exceptions to this are FPSOs which store liquids before transfer to export vessels. Similarly, where gas cannot be piped to shore, an FLNG liquefies the gas on the facility and transfers this to LNG carriers for export.

Onshore production facilities can be large and complex with multiple unit operations for separation and clean-up, or they may be very simple as for shale developments which have thousands of well sites as in the US. Offshore facilities in deep water tend to be large and complex as only larger fields are economical to produce. Some older offshore facilities in shallow water may be relatively simple.

Changes in environmental regulations over time have added complexity as routine flaring of excess gas from a primarily liquids reservoir is typically no longer permitted, and this gas must be processed and exported by pipeline or reinjected.

Generally, for the production phase whether onshore or offshore, for large developments there are similar processes: the wellheads produce into flowlines, then into manifolds ultimately connected to treatment facilities where product gas and liquids are separated, and water and sand removed. Gas may be treated and compressed to pipeline pressure. Liquids are stabilized so they are safe to store and transport without venting flammable vapors. A typical arrangement of facilities required for a production well is shown in Figure 2-10. Dehydration occurs before gas transport.

Treatment facilities for the gas phase usually include dehydration of produced gas, often acid gas removal (e.g., H_2S or CO_2), separation of condensables, and modification of heating value to allow direct injection into the commercial gas network. Mercury may need to be removed if the gas is to be used for LNG and for acceptability into most pipeline systems.

Treatment of the oil phase includes separation of water and sand and inhibiting the formation of hydrates or wax if these have the potential to build up in the export pipeline and restrict flow. Crude washing may also be required due to dissolved salt in the connate water associated with the crude along with general demulsification. The produced water phase has excess hydrocarbons removed before disposal.

Offshore unit operations are similar to onshore; however, complexity is added because of the limited space available offshore. There are many offshore facility design options as noted in Table 2-1. The facility type is determined by several factors: water depth, weather, the sea-bottom characteristics, the amount of processing required on the facility, and whether local storage is needed. An option adopted by some offshore facilities is for simple separation offshore and pipelining to shore-based facilities (with more space and lower costs) where the bulk of the treatment occurs.

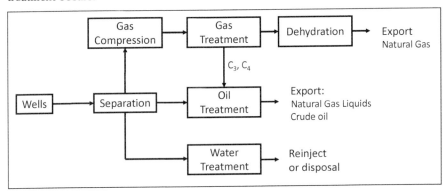

Figure 2-10. Typical arrangement of facilities for a production well

For the production phase, the BOP is replaced with a Christmas tree located on the well head for an onshore well. Offshore, the Christmas tree may be on the facility at the surface (a dry tree) or on the seabed (a wet tree or a subsea tree). The tree has multiple valves to allow production and access to the well for interventions and ultimately to kill the well if required. See Chapter 6 for additional details.

Process Safety Issues

Process safety issues during production are related to the hazardous properties of the produced hydrocarbons and other hazardous materials (see later discussion in this section). Hazardous materials include natural gas, condensate, and crude oil – these are flammable and may contain toxic H_2S. These may be present in the production equipment (e.g., separators, compressors) and also in storage tanks.

There are some important process safety issues onshore that warrant attention. Containment of toxic gases and flammable hydrocarbons from onshore storage tanks large and small is a constant process safety concern. Large facilities in harsh environments (e.g., arctic) are usually housed in modules to enable operations and maintenance in a protected environment. These are large buildings and may house the entire process other than fin-fan coolers located outside. This can lead to a potential flammable atmosphere by confining a leak that might dissipate naturally if the equipment were located outside. Forced ventilation and fire suppression, including extinguishing gases and water spray systems, are often used to protect these facilities. Although the modular building approach does impose some space constraints, this is not as limited onshore as for offshore, and thus vertical stacking of process equipment is not as common onshore. This reduces the escalation potential.

Onshore, accommodation of personnel is usually separated from production (as in Alaska). Protection of personnel in the control room would be addressed, either through location in the module or by physical protection against fire and blast hazards. Hydrocarbon storage tanks are a process safety risk, and these may also be inside a module. Onshore facilities may have frequent visits from transport drivers and others, and control of these visitors is necessary to ensure no large groupings of personnel in hazardous locations.

Offshore, inherent hazards are often greater than onshore due to the close proximity of equipment and multiple stacked decks that can permit incident escalation that are prevented onshore by suitable facility siting and layout (API 752, CCPS, 2018b). Within the industry there is a good understanding of the potential for congested equipment to lead to flame acceleration and damaging overpressures (van Wingerden, 2013). The revised design for Piper Bravo is very different than its predecessor Piper Alpha and reflects the many lessons learned from that major incident (Broadribb, 2014). Offshore designs now include fire and gas detection systems, depressurization and blowdown systems, drainage arrangements and, if appropriate, fire and blast walls. Useful guides include API 14J, Norsok S-001, and UK PFEER.

Other hazardous chemicals present onshore or offshore include methanol if used for flow assurance (flammable and toxic), glycols (e.g., TEG) for dehydration, corrosion and other inhibitors, acids for various treating, and amines (MEA, DEA, MDEA) for H_2S and carbon dioxide (CO_2) removal. H_2S separated from amine solutions can be almost pure and is very hazardous. Storage may be required for diesel fuel for emergency generators, fire pumps and jet fuel for helicopters. An environmental issue relates to naturally occurring radioactive materials (NORM) which may leach from the formation and be transported to the surface in produced water, oil and gas. These can precipitate out and form solid waste. Dumping of these locally is not permitted.

There are multiple chemicals used for well stimulation and water flood, but these normally do not pose a process safety risk. An exception is the use of hydrogen fluoride (HF) which is very toxic as well as flammable. Getting it mixed and delivered to a well, safely injected, and the returns properly handled, is a challenge.

2.5   WELL WORKOVERS AND INTERVENTIONS

During the plateau and decline phase of the well (Figure 2-1), the composition of produced fluids changes. Many reservoirs flow naturally initially because of the reservoir pressure and the gas content. Initially lighter materials may dominate, but over time the reservoir pressure generally decreases, heavier components may increase, and produced water may also increase. Also, in some cases sand can be produced. While initial flows may be free flowing, the change in reservoir characteristics to heavier materials may require some form of flow assistance (i.e., artificial lift) to produce the well – either gas lift or water injection into the reservoir to maintain reservoir pressure. Other forms of enhanced oil recovery are possible using, for example, hydrocarbons, heat, or carbon dioxide. It may also be feasible to increase production and do well maintenance through workovers or interventions.

Process Safety Issues

There are process safety implications when the well is opened for workover or intervention. Many aspects of risk are similar to well construction (see Section 2.2), including loss of well control or blowout. The likelihood may be lower if the pressure is reduced due to reservoir depletion.

2.6   DECOMMISSIONING PHASE

The end-of-life stage of a well is reached when flow rates produced are no longer economical and it is decided to abandon the well.

The proper plugging and abandonment of an onshore or offshore well is defined by regulators globally. While differences exist, they all seek to accomplish the following.

1.   Isolate and protect all subsurface freshwater zones

2. Protect all potential future commercial production zones
3. Prevent in perpetuity leaks from or into the well
4. Cut pipe to an agreed level below the surface and remove all surface equipment

When an installation has reached the end of its life, process equipment must also be safely decommissioned and removed. Onshore this is the same as decommissioning a downstream plant. Offshore, most jurisdictions now require jackets to be recovered and taken to shore. Floating installations are towed to dismantling facilities, similar to oil tankers. An early decommissioning example was the Brent Spar rig in the North Sea in 1991 which originally was to be disposed by dumping in deep water. This caused controversy as there were public concerns about potential hazardous materials remaining onboard. While these were exaggerated, the operator later decided to dismantle the rig onshore and onshore disposal is now standard in the North Sea.

Process Safety Issues

Plugging and abandoning a well can be difficult as there may have been a change of ownership and lack of accurate records of the original design and subsequent changes to the well. A process safety issue observed for onshore wells has been small leaks in the well column that lead to sustained annular casing pressure and possible groundwater contamination. In the US onshore, decommissioning rules are usually based on State regulations and offshore, BSEE provides rules for plugging wells in a Notice to Lessees (NTL 2010-G05). Vrålstad et al (2019) set out issues for offshore and common solutions in good detail. Production tubing and tools are removed and cement plugs, sometimes with supplemental mechanical plugs, are inserted at several locations up the casing. Generally, multiple plugs are required to address both the inside casing and annular spaces.

2.7 DEFINING "BARRIERS"

A major topic for process safety management covering all aspects of upstream activity is barrier definition and their ongoing management. Barriers must maintain their effectiveness through life, and this is a challenge as many are rarely used – just ready to operate if a safety demand occurs. Since the barrier topic is relevant for Chapters 4, 5, 6 and 7, it is included here as an introduction. Specific barriers are discussed later in each chapter.

Historically there have been differences in terminology regarding barriers between downstream and upstream. The terminology became more consistent with the widespread adoption of the LOPA method. A barrier, in LOPA – Layers of Protection Analysis (CCPS, 2011), is an independent protection layer (IPL) and must have specific attributes to meet this definition. Those parts of upstream similar to downstream (e.g., large gas plants and LNG facilities) also use LOPA and tend to follow the same terminology. Barriers are usually identified in a hazard identification study or PHA (CCPS, 2008a), and the need for any additional barriers for specific incident sequences may be further defined by a risk assessment, usually LOPA (CCPS, 2011) or by a full QRA (CCPS, 1999). Some confusion also exists regarding the term 'safeguard'. It is used as a collective term for any measure reducing risk whether or not it meets the criterion for an IPL. A simple phrase to remember this is "not all safeguards are IPLs, but all IPLs are safeguards".

Some formal definitions that assist in understanding the use of the term "barrier" are provided as follows.

The CCPS Glossary defines barrier and Independent Protection Layer as:

> Barrier: Anything used to control, prevent, or impede (interrupt) energy flows. Includes engineering (physical, equipment design) and administrative (procedures and work processes).

> Independent protection layer (IPL): "A device, system, or action that is capable of preventing a postulated accident sequence from proceeding to a defined, undesirable endpoint. An IPL is independent of the event that initiated the accident sequence and independent of any other IPLs. IPLs are normally identified during layer of protection analyses."

The CCPS (2018c) *Bow Ties in Risk Management* defines barrier as follows.

> A control measure or grouping of controls that on its own can prevent a threat developing into a top event (prevention barrier) or can mitigate the consequences of a top event once it has occurred (mitigation barrier). A barrier must be effective, independent, and auditable.

The term barrier is also used widely in upstream, but with different interpretations as described in the following definitions.

API 100-1 (2015b) for onshore wells provides the following barrier definition.

> "a component or practice that contributes to the total system reliability by preventing liquid or gas flow when properly installed".
> (Note: in context this refers to unplanned flows).

NORSOK D-010 provides the following definitions.

> Well barrier: Envelope of one or several dependent barrier elements preventing fluids or gases from flowing unintentionally from the formation into another formation or to surface.

> Well barrier element: Object that alone cannot prevent flow from one side to the other side of itself.

IOGP 415 defines barrier as follows.

> A risk control that seeks to prevent unintended events from occurring or prevent escalation of events into incidents with harmful consequences. Hardware and human barriers are put in place to prevent a specific threat

or cause of a hazard release event, or to reduce the potential consequences if barriers have failed and an event has occurred. Both hardware and human barriers are supported by the processes and procedures contained within the Management System Elements.

Comparing the definitions provided above, the differences highlighted below illustrate why it is helpful to propose a single definition for the purposes of this concept book.

- The API 100-1 definition is not specific and allows a barrier to be what D-010 terms a barrier element contributing to but not necessarily forming a complete barrier in itself.

- NORSOK D-010 barriers are generally one or more hardware / physical elements which collectively define the barrier for a well system. It does not include important human barriers.

- The CCPS definition for IPL and the bow tie definitions are close to IOGP and state that a barrier must be capable of terminating an event sequence on its own and the barrier must be effective, independent, and auditable. It allows for both hardware and human barriers.

In this book the CCPS bow tie barrier definition is used.

Barrier: A control measure or grouping of control elements that on its own can prevent a threat developing into a top event (prevention barrier) or can mitigate the consequences of a top event once it has occurred (mitigation barrier). A barrier must be effective, independent, and auditable.

Additional concepts included in the definition are explained below.

- Effective – fully prevents the unintended event or mitigates (at least to a worthwhile degree) the specified undesired consequence(s).

- Independent – functions independently of the initiating event and the design or operation of any other barrier, without having a common sensor or logic in common with another barrier.

- Auditable – evidence exists that the barrier is present and will reliably function as intended (sometimes referred to as verifiable).

Upstream facilities usually have many barriers deployed for multiple major hazard scenarios. Bow tie diagrams provide a holistic graphic understanding of these barriers more clearly than simple lists. Bow tie diagrams have been deployed to enhance communication and understanding for everyone, from front line operators to senior management. The bow tie approach to risk management (CCPS, 2018c) differentiates prevention barriers from mitigation barriers. A simplified bow tie diagram is shown in Figure 2-11. A prevention barrier appears on the left-hand side before the top event (i.e. a loss of containment) and mitigation barriers on the right-hand side. Prevention barriers must in principle be capable of terminating an event sequence, whereas mitigation barriers may permit the sequence to continue but in a mitigated form. Only a single barrier for each threat and consequence is shown in this illustrative figure on the prevention and mitigation sides, but in reality, there can be several. The top event is the loss of control over the hazard, most commonly an LOPC event.

The differentiation of barriers and degradation controls was discussed in the concept book *Bow Ties in Risk Management* (CCPS, 2018c). Bow tie diagram pathways drawn incorrectly and showing multiple barrier elements, rather than whole barriers meeting the IPL standard, can make personnel discount that sequence as very unlikely due to the appearance of many apparent 'barriers'. An example of this is for a loss of well control bow tie – there are two primary prevention barriers (the mud column, the well pressure containment system/ BOP), with the well pressure containment system and BOP not being independent of each other. But some bow ties show many additional prevention barriers. These extra barriers are all useful degradation controls, as they support or enhance the operation of the two primary barriers, but they are not full barriers themselves. Examples of these are the multiple seals and connection isolations associated with casing and the BOP, and the electro-mechanical system controlling the BOP. Drawing the bow tie correctly and showing only two barriers on the main pathway highlights to personnel the necessity to manage these well. The full diagram shows all the degradation controls supporting the primary barriers so that those managing the system fully understand its operation and potential failure modes. Drawn this way, the bow tie diagram does not convey a false picture of many barriers when there are only two.

Bow tie diagrams can provide additional information such as expected reliability, the barrier owner, and relevant documentation. CCPS (2018c) provides a full list of such information.

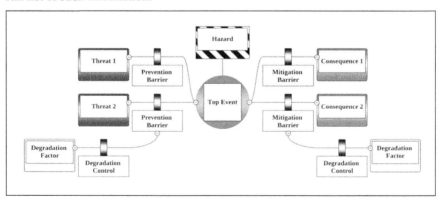

Figure 2-11. Bow tie model

Barriers can fail and are not perfectly reliable and this is often explained using Reason's swiss cheese analogy (e.g., IOGP, 2018a) – a slice of cheese representing the barrier and the holes representing its potential failure on demand. In principle, therefore, multiple barriers are needed to give confidence that a threat pathway is always terminated or adequately mitigated. Bow ties can be used to identify and share important barriers during design and operation, and if an incident occurs, to explain which barriers failed.

2.8   OVERVIEW OF INTERNATIONAL REGULATIONS

Process safety regulations exist for most offshore regions. A trend has been to separate process safety and environmental regulators from government functions promoting the development of offshore resources. Example safety regulators (some with environmental responsibilities as well) include the following.

- BSEE (Bureau for Safety and Environmental Enforcement) in US
- HSE (Health and Safety Executive) in UK
- PSA (Petroleum Safety Authority) in Norway
- ANP (National Agency of Oil, Gas and Biofuels) in Brazil
- C-NLOPB (Canada-Newfoundland and Labrador Offshore Petroleum Board) in Newfoundland, Canada
- NOPSEMA (National Offshore Petroleum Safety and Environmental Management Authority) in Australia

These and other regulators meet periodically and publish various technical studies and incident statistics at the International Regulators Forum website (irfoffshoresafety.com).

There are two broad approaches to regulation: prescriptive and goal-based. This differentiation, and their historical development, are discussed fully for downstream applications in CCPS (2009). Historically, all safety regulations were prescriptive, usually developed in response to an incident, and where the regulation specified the nature of the safety remedy required (Broadribb, 2017). An example of a predominately prescriptive regulatory approach for upstream is seen in the US, although elements of goal-based appear in the SEMS regulation it also spells out exactly what must appear in the management system and thus is partly prescriptive. Examples of the goal-based regulatory approach for upstream are given by the UK, Norway and Australia – all for offshore. Canada, in 2020, is in transition from a prescriptive to a goal-based approach. These approaches are not pure; goal-based regulations include some prescription and prescriptive regulations include some goal-based aspects.

Major downstream incidents in the US (e.g., vapor cloud explosion in Pasadena, Texas in 1989) resulted in OSHA developing the Process Safety Management (PSM) Regulation (OSHA 1910.119) which was phased in over several years commencing in 1992. This was based extensively on API RP 750. This applies mainly to downstream refining, chemical and petrochemical plants. It also covers some upstream facilities, such as large onshore treatment facilities and gas plants, but not well construction. Coverage is based on threshold amounts of nominated hazardous chemicals. Other incidents, including Bhopal in 1984 (Less, 2012), led the EPA to issue the Risk Management Plan Regulations, which are similar to PSM, but with a focus on offsite impacts. As well as a formal process safety management system, the EPA requires that the facility carry out an offsite consequence analysis to predict worst case hazard zones and to report on 5-year incident histories. This regulation also applies to larger scale upstream treatment plants based on their maximum inventory, similar to OSHA PSM.

In the prescriptive approach, the company and the regulator determine, by a review of documents or by inspection, whether the regulatory requirement has been implemented. This process ensures that good solutions are adopted. It does require that the regulator or industry body develops and keeps up to date the required solutions. Maintaining current solutions is demanding in industries such as deepwater offshore, which have complex and changing designs. A potential issue with prescriptive requirements is that once the remedy is implemented, then a culture of compliance may develop and not consider other solutions, even if they could potentially provide additional risk reduction.

The EU approach for downstream is goal-based. Two downstream major incidents occurred in Europe in the 1970s (Flixborough and Seveso) and these along with the Bhopal incident in 1984 (all described in Lees, 2012 and CCPS, 2008b) resulted in the EU developing the Seveso Directive, a goal-based process safety regulation. A summary of the current requirements is provided on the UK HSE website (see references). This requires a formal process safety management system and a risk assessment as part of a safety case document. In the goal-based approach the process safety objective is defined, but the actual solution is left to the company. The safety objective is often ALARP – As Low As Reasonably Practicable. For ALARP, additional safety measures to reduce risk should be assessed and implemented so long as the measure is practical, considering trouble, time, and money. The company must demonstrate to the regulator that the solution adopted is adequate for the potential risk.

In the upstream domain, the Piper Alpha disaster in the UK sector of the North Sea in 1988 resulted in 167 fatalities (IChemE, 2018 and Oil & Gas UK, 2013). The subsequent Cullen Inquiry (HSE, 1990) recommended, amongst other things, a safety case approach for offshore. This requires a safety case or for floating drill rigs, an IADC-style HSE Case. The safety case includes a facility description, major hazards identification and risk assessment, the process safety and environmental management system, the technical solutions (addressing key barriers – their performance standards, how they are maintained, and verification), and the emergency response actions/procedures. The UK regulations are implemented in a family of regulations led by safety case which is goal-based but including more prescriptive regulations such as PFEER – Prevention of Fire and Explosion and Emergency Response. The EU adopted a directive addressing the safety of offshore oil and gas operations (2013/30/EU) in 2013. The onshore upstream industry in Europe is small and it is covered by the Seveso Directive.

The US offshore industry had no legislated process safety management system requirements until after the Deepwater Horizon disaster in 2010. However, beginning in the early 1990s, many larger, integrated oil and gas companies had deployed their own internal process safety management systems. After Deepwater Horizon, the USA made API RP 75 mandatory in a regulation titled Safety and Environmental Management Systems (SEMS – 30 CFR Part 250 Subpart S). BSEE updated this after a short period to include some additional aspects titled SEMS II. The upstream onshore industry, as previously noted, has PSM and RMP requirements for large facilities, but not for smaller ones and drill sites. Some states require safety management systems, but this is not uniform.

A direct detailed comparison of international approaches related to process safety management systems is difficult as the regulations are structured differently. However, in many aspects, regulations cover similar topics. For example, the majority of regulations require workforce involvement, process knowledge management, hazard identification, operating procedures, safe work practices, asset integrity, and emergency management.

Some important differences include the following.

- Leadership and accountability has always been required under safety case approaches. These are not a requirement of OSHA 1910. The recent 4th edition of API RP 75 includes leadership as does BSEE in its safety culture guidelines. COS has issued a leadership site engagement guideline (COS, 2014) and a safety culture guideline (COS, 2018). The UK has emphasized leadership engagement for a long time (e.g. HSE R2P2 (2019b) updated document). The Norwegian PSA has also emphasized leadership and accountability in its offshore process safety system, which was reviewed in detail after the Deepwater Horizon event and found no need to change.

- RBPS recommends hazard identification and risk analysis. Both OSHA and SEMS specify only hazard identification. BSEE has made the risk assessment aspect essentially a requirement. Usually this is achieved by a risk matrix approach in conjunction with the hazard identification. The Norway PSA requires a full QRA. The UK falls in between.

- Some differences also exist on management review and continual improvement (aka continuous improvement).

This book does not promote one approach or the other. Either approach or any combination can be successful with the support of competent personnel (company, regulator, and other stakeholders) and engaged leadership.

3

Overview of Risk Based Process Safety (RBPS)

3.1   BACKGROUND

Formal management system approaches to control operations and finance have been well established in business for many years. Quality management systems were first created by pioneers such as Deming, Juran and Shewhart (Deming, 1993) including the PDCA Loop (Plan – Do – Check – Act), and later codified into formal standards (initially BS5750 in 1979 and later adopted as ISO 9001). The next step was for formal environmental management systems (initially BS7750 in 1992 and later adopted as ISO 14001) and safety management systems (ISO 45001). This latter standard has a greater focus on occupational safety rather than process safety.

Process safety management systems have been recognized as fundamental to achieving process safety since at least the late 1980s. Many companies had their own internal systems and the best of these provided inputs to early published examples including Responsible Care (initially in Canada in 1984 and adopted by the US chemical industry in 1988) and from CCPS (1991). Details are available from the International Council of Chemical Associations website (see References for web address). API was also active and issued recommended practices for downstream (RP750 in 1990) and for upstream (RP 75 in 1993).

CCPS issued the *Guidelines for Risk Based Process Safety* – RBPS in 2007 (CCPS, 2007a). The inclusion of risk assessment is believed to improve process safety by better addressing high consequence / low frequency events. The aim of RBPS is to ensure that risk management of process safety is embedded throughout the process safety management system.

RBPS is voluntary industry guidance. Implementing RBPS extends beyond the compliance requirements of for example OSHA PSM onshore, and incorporates current practices for managing risk. RBPS applies equally to downstream and upstream facilities.

3.2   RBPS SUMMARY

Throughout this book the various discussion topics and case studies are related to RBPS. Consequently, it is important to have a general understanding of Risk Based Process Safety. A brief overview of how each RBPS element impacts upstream is provided below and greater detail on this relationship is provided in following chapters.

The RBPS structure is composed of 20 management system elements grouped under four pillars.

1. Commit to Process Safety
2. Understand Hazards and Risk
3. Manage Risk
4. Learn from Experience

RBPS is not a prescriptive management system. It sets out a structure that addresses key topics. Companies use their own structure in meeting their objectives, conforming to regulations, and adding any additional topics from RBPS. These additions do not impact audit systems such as those issued by the Center for Offshore Safety (COS, 2014).

The pillar structure including the management system elements is provided in Figure 3-1. Following this, each pillar and its elements are described briefly along with example incidents illustrating the application of RBPS. The full specification is available in CCPS (2007a).

Pillar: Commit to Process Safety
- Process safety culture
- Comply with standards
- Process safety competency
- Workforce involvement
- Stakeholder outreach

Pillar: Understand Hazards and Risk
- Process knowledge management
- Hazard identification and risk analysis

Pillar: Manage Risk
- Operating procedures
- Safe work practices
- Asset integrity and reliability
- Contractor management
- Training and performance assessment
- Management of change
- Operational readiness
- Conduct of operations
- Emergency management

Pillar: Learn from Experience
- Incident investigation
- Measurement and metrics
- Auditing
- Management review and continuous improvement

Figure 3-1. RBPS structure

3.2.1 Pillar: Commit to Process Safety

The aim for the first pillar is to ensure that the foundation for process safety is in place and embedded throughout the organization.

RBPS Element 1: Process Safety Culture

This element describes a positive environment where employees at all levels are committed to process safety. It starts at the highest levels of the organization and is shared by all. Process safety leaders nurture this process. Process safety culture is differentiated from occupational safety culture as it addresses less frequent major incident prevention cultures as well as occupational safety.

This element highlights the necessary role of leadership engagement to drive the process. Safety culture should not be thought of as a passive outcome of a specific work environment; rather it is something that is managed and improved.

Example Incident: Deepwater Horizon

The Deepwater Horizon National Commission identified poor process safety culture as a leading cause of that incident, while simultaneously having a positive occupational safety culture and achievement of excellent performance with traditional safety indicators. High drilling costs, delays, and a desire to move onto the next task led to poor decision making and discounting of danger signs. Similar issues have been apparent in the downstream industry as well. The US National Academies (2016) reviewed how to strengthen safety culture in the offshore oil and gas industry and they endorsed the BSEE nine characteristics of a positive safety culture. The UK HSE has also addressed this topic in multiple publications, including Reducing Error and Influencing Behaviour (HSE, 1999).

RBPS Application

Process Safety Culture provides an overview and suggested means on how to improve process safety culture.

RBPS Element 2: Compliance with Standards

Organizations should comply with applicable regulations, standards, codes, and other requirements issued by regulators and consensus standards organizations. These requirements may need interpretation and implementation guidance. The element also includes proactive development activities for corporate, consensus, and governmental standards.

RBPS Element 3: Process Safety Competency

This element addresses skills and resources that a company should have in the right places to manage its process safety hazards. It includes verification that the company collectively has these skills and resources and that this information is applied in succession planning and management of organizational change.

Example Topic: Standards

Some standards apply to both upstream and downstream (e.g., ASME VIII, API 520), but there are also important differences. API has issued 50 to 100 recommended practices for upstream onshore and offshore, and these differ from most downstream standards. In the US, companies must define the standards they intend to use (OSHA RAGAGEP regulation – Recognized and Generally Accepted Good Engineering Practice), and they are accountable against these. In the goal-based approach, companies need to develop a safety case, which includes amongst other matters the standards used, and the safety case and specific national regulations become the requirements for the company.

RBPS Application

Compliance with Standards sets out suggested means to comply with the range of relevant standards, codes and regulations. Internal specialists may be required to interpret the application of these to specific upstream activities. Finally, participation in standards committees and providing regulation feedback is an essential activity to ensure that these documents are up to date and cover engineering issues properly.

As several investigations have shown, excellent performance in occupational safety does not guarantee similar performance in process safety. Personnel may have a good understanding of the precursors to occupational incidents and the barriers and behaviors that prevent these, but not necessarily the same level of understanding/knowledge for more complex and rarer process safety events.

Companies need to assure themselves that personnel at all levels understand how to apply process safety principles. This competency requirement applies across the life cycle (exploration, well construction, design, production, and abandonment), including tasks that are rarely executed. A system for verification of this competency is necessary.

RBPS Element 4: Workforce Involvement

This element addresses the need for broad involvement of operating and maintenance personnel in process safety activities to make sure that lessons learned by the people closest to the process are considered and addressed.

Workforce involvement in process safety reviews (e.g., risk assessments, management of change, etc.) ensures their knowledge of potential problems and operation of key safety systems is included and considered in potential risk reduction enhancements. Their involvement also helps build an understanding in personnel of major hazards and how barriers are deployed to make these safe. Most offshore regulators require workforce involvement for process safety.

RBPS Element 5: Stakeholder Outreach

This element covers activities with the community, contractors and nearby facilities to help neighbors, outside responders, regulators, and the general public understand the asset's hazards and potential emergency scenarios.

This activity may be coordinated by local regulators for smaller onshore upstream facilities, but usually is driven by the company for larger integrated facilities. It also applies offshore, but the stakeholders may be different. Large onshore facilities may be covered by national regulations and these require sharing of the possible hazard zones and emergency plans. Where there are joint owners, the outreach can resolve any differences in owner process safety policies.

Example Topic: Stakeholder Communications

A shale development in Colorado was located very close to suburban housing and caused major concerns to the community. The local regulator, in that case the City and County of Broomfield, worked with the company to hold a series of public meetings to address community concerns. A set of safety and environmental management best practices was agreed with the company, and this formed the basis of regulatory controls. The communication process was better than the company could have achieved on its own due to the strong emotions raised in the community and the need for a neutral party to facilitate contacts.

RBPS Application

Stakeholder Outreach directly addresses this issue and allows for alternative means for communication.

3.2.2 Pillar: Understand Hazards and Risk

This pillar addresses process knowledge management and the identification of hazards and management of risks. Process knowledge must be readily available and kept up to date. Risk management is an extension of some regulatory requirements, but in practice most companies conduct hazard identification and add risk ranking as part of the analysis. Inherently safer design considerations are applied here. RBPS also notes that companies may choose to go beyond qualitative risk ranking to some form of quantitative risk assessment (LOPA or QRA).

RBPS Element 6: Process Knowledge Management

Process knowledge management is the assembly and management of information needed to perform process safety activities. It includes verification of the accuracy of this information and confirmation that this information is kept up to date. This information must be readily available to those who need it to safely perform their jobs.

Upstream facilities, whether onshore or offshore, are often remote and knowledge and documentation of importance to personnel may be located in central offices. Access to this knowledge and documentation should be arranged and involve regular meetings or communication links. An incident where the importance of access to this knowledge was highlighted is the Longford incident (see box below) where the important knowledge on brittle fracture susceptibility was not available locally at the time. Important knowledge includes incident lessons and datasets, updated engineering standards, equipment drawings and specifications, operational experience and upsets, and new or updated process safety tools.

During engineering projects (see Chapter 7) teams can change at each stage, and it is important that relevant process safety knowledge is transmitted along with the design.

Example Incident: Longford

A well-known example of a depleting well leading to a process safety incident is the Longford gas plant fire in Australia (Hopkins, 2000). The field, located in the Bass Strait, was gradually producing more heavy ends. A separation column in the plant no longer could function effectively and ultimately during an upset allowed natural gas liquids (NGLs) to enter part of the plant not designed for it. This reduced the temperature within a heat exchanger to -45°C causing the exchanger to be completely frozen. Operators tried to diagnose the issue, but process safety personnel who might have known about embrittlement had been transferred to Melbourne and were not readily available to assist with the diagnosis and warn of the dangers of cold temperature embrittlement. The operators introduced hot oil to unfreeze the exchanger, but the resulting thermal stress led to a brittle fracture rupturing the vessel causing injuries and fatalities for those nearby. A long-lasting fire ensued and ultimately gas supply for the entire state had to be terminated for several days resulting in significant economic losses.

RBPS Application

Management of Change – MOC would have identified the change in processing conditions that might allow NGLs to reach parts of the process not intended for this service. MOC also includes organizational changes and this would have addressed the significance of relocating process safety experts remote from the facility.

Hazard Identification – As part of the MOC process, a HAZID would have identified that the change in incoming fluids could allow NGLs to reach parts of the facility not intended for this service. This could lead to flash vaporization and cold temperatures in places with mild steel material.

Process Safety Knowledge – Since there was the potential for NGLs to vaporize and drop temperatures to -40°C, this could cause embrittlement in mild steel. Personnel should have been made aware of this threat to ensure their operational responses would not impose major thermal stresses if this occurred.

Note: some additional issues for Longford are presented in Chapter 5.

RBPS Element 7: Hazard Identification and Risk Analysis

Hazard Identification and Risk Analysis (HIRA) are complementary activities initially identifying process safety hazards and their potential consequences and later estimating the scenario risks. HIRA includes recommendations to reduce or eliminate hazards, reduce potential consequences, or reduce frequency of occurrence. Analysis may be qualitative or quantitative depending on the level of risk. HIRA is a core process safety activity.

HIRA analyses vary from simple to complex. In addition to basic topics such as identifying responses to upsets, potential leak scenarios, important barriers and integrity, it must also take into account extreme and remote environments, reservoir uncertainties, and compounds that affect production (e.g., waxes or radioactive materials). It also must consider the potential exposures to people (public can be nearby onshore, or personnel accommodations may be located next to the facility offshore or at remote onshore facilities), and to the environment and the asset.

Many different tools are used for HIRA analyses. These range from simple checklists, through What-If and HAZOP, to more complex LOPA, QRA, fire hazard analysis and explosion studies. Inherent safety methods and functional safety assessments fall within HIRA.

Example Incident: Piper Alpha

The Piper Alpha incident in 1988 resulted from deficiencies in the RBPS risk management pillar, but also had problems related to safe work practices (defective work permit system) and emergency management (no safe place for refuge and backup control if the control room was disabled). The facility was modified to meet updated environmental regulations. Initially it just handled liquids and gas was flared, but to avoid this a gas compression and export module was added. The layout was not ideal and resulted in major process facilities being too close to the control room. When the event occurred, the control room was quickly disabled and the explosion event escalated to multiple pool fires and later a major jet fire. Personnel congregated in the accommodation module and perished there due to smoke inhalation. At the time there was no requirement for detailed risk assessment to track an initial event and how this might escalate to involve other modules and release more hydrocarbons. The Cullen Inquiry recommended that a QRA be carried out addressing such risks and ensure safety systems could prevent the escalation.

RBPS Application

The hazards of high-pressure hydrocarbons were reasonably well known at the time, but not the risk of escalation. *Hazard Identification and Risk Analysis* sets out the means to take a hazard identification and extend this with a risk assessment addressing possible escalations. Escalation is more important upstream, especially offshore, where spacing is limited with the small footprint available. The Piper 25 conference (Oil & Gas UK, 2013) and Broadribb (2014) outline the major learnings and modifications since 1988.

It is a truism that what has not been identified cannot be prevented or mitigated. HIRA activities should be translated into a risk register and action tracking system for any needed follow-up activities. This is to ensure that no identified issue is inadvertently neglected during progression through the design phases.

3.2.3 Pillar: Manage Risk

This pillar addresses many important topics for operational safety and management of risks. These include operating procedures, safe work practices, contractor management, training, operational readiness and conduct of operations. This pillar also addresses asset integrity, management of change, and emergency management. All these topics are important to process safety for upstream.

RBPS Element 8: Operating Procedures

These are written instructions for an activity that describe how the operation is to be carried out safely, explaining the consequences of deviation from procedures, identifying key safeguards, and addressing special situations and emergencies.

Operating procedures have improved substantially from the past approach of simply taking start-up procedures from the design contractor. Now procedures are designed with operating personnel engagement, are periodically updated based on feedback and any modifications, and use modern layouts with graphics and photographs to convey key safety messages. Risks from deviations are highlighted – e.g., if equipment purging is required before start-up, the procedure should highlight safety risks with shorter duration purging. Barrier management is an important aspect of process safety and the procedures should highlight relevant barriers potentially affected by the procedure.

RBPS Element 9: Safe Work Practices

Safe work practices are requirements established to control hazards and are used to safely operate, maintain, and repair equipment and conduct specific types of work. They include control of work (job safety analysis (JSA), permits and oversight), breaking containment, energy isolation, SIMOPS (see SIMOPS discussion in Chapter 5) and other activities. These practices are used when developing detailed procedures, ensuring that requirements are met and the appropriate safeguards have been or will be implemented for the work.

In upstream facilities, there can be several parties involved in work – the owner and its contractors. Interface documents dictate what safe work practices are used and specify who approves the work.

RBPS Element 10: Asset Integrity and Reliability

Asset integrity and reliability activities ensure that important equipment remains suitable for its intended purpose throughout its service lifetime. This includes proper selection of materials; inspection, testing, and preventative maintenance; and design for maintainability. During the design stage, potential asset integrity problems can be anticipated and significantly mitigated.

Example Incident: P-36 off Brazil

The P-36 FPU explosion and sinking event offshore Brazil (Barusco, 2002) is an example where hazard identification did not address a possible problem. The event was initiated by a pressure burst of a drain tank in one of the leg columns. The cause was an incorrect isolation of the vessel during long-term maintenance as one connection was isolated using only a valve, not a blind. That permitted drain water containing hydrocarbons to seep past the valve into the vessel. The relief had also been isolated, and this meant that the internal pressure increased as the liquid volume built-up compressing the trapped vapor space above. The vessel ultimately burst and released flammable hydrocarbons into the column space.

The emergency response team was not aware that any hydrocarbons would be present, and they accidentally ignited the flammable mixture. This killed several members and ruptured the main cooling water supply line that ultimately sank the vessel. The team and operations personnel did not recognize the potential presence of the hydrocarbon hazard.

RBPS Application

Safe Work Practices directly addresses the need for safe work practices to be correctly followed. In this case it would refer to Isolation Procedures.

HIRA (Hazard Identification and Risk Analysis) should have identified the potential for hydrocarbons to be present in the column area due to the direct connection of the process into the drain tank. This should have been communicated as part of the emergency response procedures.

Equipment and control systems can be affected by harsh onshore and offshore environments. Some equipment can be hard to inspect, particularly on offshore installations. Offshore and remote onshore installations may have accommodation limits that reduce the availability of visiting personnel to perform integrity tasks. In the EU offshore, and for many companies onshore and offshore, there is a focus on safety critical elements and achieving performance standards.

Upstream reservoirs decline with time and new wells may be drilled or stimulation activities with potentially corrosive chemicals employed. This may bring asset integrity issues. Similarly, many upstream facilities are operating beyond their intended design life and are managing aging issues.

RBPS Element 11: Contractor Management

Contractor management is a system of controls to ensure that contracted products and services support (1) safe operations and (2) the company's process safety and occupational safety performance goals. It includes the selection, acquisition, use, and monitoring of contracted products and services. These controls ensure that contract workers perform their jobs safely, and that contracted products and services do not add to or increase safety risks.

Example Incident: Ocean Ranger

The Ocean Ranger semi-submersible sank offshore Canada in 1982 due to heavy weather and strong waves. A porthole failed and this allowed seawater from waves up to 65 feet (20 m) high to reach the vital ballast water control system. The seawater soon caused the control system to operate unpredictably and a list developed. The rig was abandoned but none of the 84 crew members survived.

RBPS Application

Asset Integrity and Reliability emphasizes that critical operational and safety systems integrity be maintained – and this is more than simple reliability estimations for the system itself. The system should be protected against failures caused by other systems or the external environment.

Contractors are prominent in operations, maintenance, well construction, workover, intervention and decommissioning activities. They have specialist knowledge and equipment to enable challenging tasks to be performed safely and efficiently. It is necessary to align the process safety program of the company with its contractors to ensure that all aspects are addressed and that everyone knows their responsibilities. API and IADC provide guidance for interface agreements that help to formalize this process.

RBPS Element 12: Training and Performance Assurance

Training refers to practical instruction on the job and task requirements and methods for operators, maintenance workers, supervisors, engineers, leaders, and process safety professionals. Performance assurance verifies that the trained skills are being practiced proficiently.

Upstream work is challenging, and a high degree of skill is needed to perform tasks correctly. Numerous upstream incidents identify weaknesses in training and job execution as underlying causes, including, for example, the Black Elk incident

Example Incident: Black Elk

The Black Elk incident in 2012 offshore in the Gulf of Mexico was an example of poor contractor management and communications. Multiple contractors were working on the platform. The incident involved welding on a line into a tank system. This caused an internal explosion that ejected three tanks off their bases, resulting in three fatalities and injuries to others. The incident was investigated by BSEE (2013). They identified issues with contractor management – poor communication and culture, poor safe work practices, improper hot work control, and failure to actuate stop work authority, amongst other issues.

RBPS Application

Contractor Management aims to ensure contractors are aware of all hazards and have training in required safe work practices. This is key in upstream operations where many different contractors interface on a daily basis.

described above. More formal training needs analysis helps underpin a necessary training program. This should include process safety hazards and how to participate in or interpret risk analysis studies, as appropriate.

Formal testing of knowledge and skills is an important part of this element to assure that participants have understood the material. It includes on-the-job task verification.

RBPS Element 13: Management of Change

Management of Change (MOC) is a system to identify, review, and approve modifications to equipment, procedures, raw materials, processing conditions, and people or the organization ((CCPS, 2013b), other than replacement in kind. This helps ensure that changes are properly assessed (for example, for potential safety risks), authorized, documented, and communicated to affected workers. Documentation includes changes to drawings, operating and maintenance procedures, training material, and process safety documentation. The objective is to prevent or mitigate incidents prompted by unmanaged change.

Many past incidents have been due to changes that were not properly assessed, and which defeated existing safeguards or introduced new hazards. For example, the Piper Alpha platform was modified after start-up to include gas recovery, but without sufficient risk assessment of the higher risks this introduced (Broadribb, 2014).

The MOC is a formal process involving similar tools to the initial hazard identification and risk assessment. A newer aspect of MOC is the recognition that organizational change can also create process safety issues and a specific Management of Organizational Change (MOOC) procedure has been developed (CCPS, 2013b).

Well operations are subject to changes as the process of drilling is dynamic. Unexpected geological conditions may be encountered that make the initial well plan invalid and require an update. This should be the subject of an MOC review to ensure new hazards are not introduced by the change.

RBPS Element 14: Operational Readiness

Operational readiness evaluates the process before start-up to ensure the process can be safely started. It applies to restart of facilities after being shut down or idled as well as after process changes and maintenance, and to start-up of new facilities.

An important aspect is to verify that all barriers identified in design reviews and captured in a risk register and action tracking system have been implemented and/or any outstanding actions are approved for later close-out.

An aspect for upstream in onshore remote locations and offshore in GOM is the need to shut down and de-man during severe weather events (hurricanes) and hence to restart more frequently than is the normal for process facilities. Similarly, startup of not normally manned operations requires special attention.

RBPS Element 15: Conduct of Operations

Conduct of operations (CCPS, 2011) is the means by which management and operational tasks are carried out in a deliberate, consistent, and structured manner. Managers and co-workers ensure tasks are carried out correctly and prevent deviations from expected performance.

This element aims to ensure that there is operational discipline; in other words, that operating practices are applied correctly and fully every time. In doing this, normalization of deviance can be avoided (i.e., hazardous shortcuts that succeed once and become the standard way of doing the activity). This is addressed in CCPS (2018e). It also identifies an operational envelope beyond which procedures no longer apply. Personnel and contractors must recognize these conditions and either stop the activity or call in more experienced personnel to address the deviation. A challenge for upstream is ensuring conduct of operations given a high personnel turnover due to heavy use of contractors and operations in remote locations.

SIMOPS activities are also covered by this element to identify interactions between the activities that might create hazards not addressed individually by either.

RBPS Element 16: Emergency Management

This element addresses plans for possible emergencies. These define actions in an emergency, resources to execute those actions, practice drills, post drill reviews and continual improvement, training or informing employees, contractors, and neighbors and local regulators, and communications with other stakeholders in the event an incident does occur.

Many upstream facilities onshore and all offshore facilities are remote. This limits immediate aid, although offshore there is now a common requirement for support vessels with firefighting capability. Personnel escape and evacuation is generally straightforward for onshore facilities as there are usually multiple egress routes and no requirement for support equipment other than standard PPE or breathing sets if toxics are present. Offshore, the situation is more difficult as escapeways are limited and lifeboats are usually required for evacuation, as it is unsafe for the personnel to jump into the sea from height and potentially into harsh environments.

Firefighting philosophies can be different upstream. Onshore remote not normally manned facilities may have an 'isolate and allow to burn out' strategy, whereas large manned facilities may have full firefighting capability. Mutual aid schemes, common in the downstream industry, may be challenging at onshore remote or offshore locations.

Loss of well control issues are unique to upstream. It requires drillers to be continuously alert to identify a kick event which may commence very subtly with few obvious signals. Dealing with a kick requires effective emergency response plans, procedures and resources to ensure that this does not result in a loss of well control. Section 4.3.4 gives details on kick detection.

> **Example Incident: Piper Alpha**
>
> The Piper Alpha incident in 1988 is a rich source of lessons learned (see prior highlight box under the element of Hazard Identification and Risk Analysis). This aspect of the incident relates to how the control room was disabled immediately by the first explosion incident. This was where emergency management should have occurred. Due to fires and smoke, personnel retreated to the accommodation module to await further instructions and evacuation. The accommodation was not smoke resistant and most of the fatalities occurred from smoke inhalation in that location. A few personnel jumped into the sea and survived the fall and were rescued. Neighboring facilities continued to pump hydrocarbons towards the Piper Platform even after they were aware it was on fire as they did not have permission from shore management to stop the flow.
>
> Several important lessons were learned. The emergency management center should be protected against potential incidents and a backup control center should exist in a separate and distanced location to enable emergency management if the first center is lost. There should be temporary safe refuges for personnel to escape to prior to evacuation. These locations must be protected against fire and smoke and allow transit to evacuation points – usually the lifeboats. Further, emergency response should be delegated to operational personnel on adjacent facilities – a type of stop work authority, without the need for shore-based management approval.
>
> **RBPS Application**
>
> *Emergency Management* sets out a comprehensive approach to emergency management which ensures that all aspects of emergency preparedness are carefully assessed, and that emergency response is practiced so that everyone knows what needs to happen.

3.2.4 Pillar: Learn from Experience

This pillar addresses how to learn from experience – from incidents and near misses, from leading and lagging metrics, and from audit reports. The aim from all of these is to identify deeper systemic causes and to implement corrective actions that solve the wider issue, not just the specific instance. The final element of management review leads to continual improvement. This goes beyond prescriptive requirements. Good process safety should seek improvements by learning from experience, going beyond compliance which can degrade into a tick-the-box mentality.

RBPS Element 17: Incident Investigation

Incident investigation is the process of reporting, tracking, and investigating incidents and near misses to identify root causes so that corrective actions are taken, trends are identified, and learnings are communicated to appropriate stakeholders.

Example Issue: Learning from Experience

Trevor Kletz, a renowned safety expert, told a story at a Mary Kay O'Connor process safety conference of how after an incident his manager asked him to drop everything and write a full report. He replied that there was no need as the report was already in the files – from prior incidents. There are multiple examples of blowout and other upstream incidents which could have been prevented if lessons from the past had been learned (see box under Incident Investigation section).

RBPS Application

The entire pillar of *Learn from Experience* is designed to ensure that lessons from past incidents are systematically integrated into existing management systems and applied to future operations and assets so that companies do not have to rely on corporate memory to prevent the recurrence of similar incidents.

It is essential that the learnings are shared with management and workers at the facilities through regular safety talks or in formal training as they have the most to gain from the information and have the ability to ensure the learnings drive the necessary changes. The management review (see Element 20), normally annually, reviews incident investigations and ensures the recommendations are implemented, including actions to address broader systemic concerns.

An important aspect for upstream is that many facilities are in remote locations and evidence from incidents needs to be preserved until the investigation can occur.

Example Issue: Learning from Incidents

Many investigations of upstream incidents focus on equipment failures and human errors as causation. They do not delve into underlying causes which are usually related to deficient management systems or organizational factors. The UK HSE now requires that incident investigations there include an assessment of Human and Organizational Factors (HOF) as underlying root causes.

There has been a lack of learning from some upstream incidents. For example, the Montara blowout event in August 2009 had many similar elements of causation to the Macondo blowout in April 2010 (National Commission, 2011) and to the onshore Pryor Trust blowout in January 2018 (CSB, 2019). An issue related to learning is that the first event was in another country and the two recent US blowouts were offshore and onshore respectively. These factors do make lesson sharing more difficult, but no less valuable.

RBPS Application

Incident Investigation sets out a comprehensive approach to incident investigation that drives to root causes and addresses the effectiveness of barriers, including human and organizational factors. It also aims to ensure that lessons are learned through effective communication.

Personnel and contractors can be multilingual, and communication requires extra effort.

RBPS Element 18: Measurement and Metrics

Current trends are an important indicator of process safety performance. This element addresses leading and lagging indicators of process safety performance, including incident and near-miss rates as well as metrics that show how well key process safety elements are being performed. Guidance is available on suitable metrics from API for downstream (API 754) and IOGP for upstream (IOGP 456). Upstream-specific metrics are defined such as loss of well control events or maritime incidents.

Contractors perform much of the work in upstream, so it is important that their process safety performance is included in the company collected metrics.

API 754 and IOGP 456 are closely aligned and have four tiers of process safety events with Tiers 1 and 2 being actual losses of containment, fatality or multiple injury events. Tiers 3 or 4 address demands on safety systems, near miss events, delayed inspections and maintenance, and deficient safety management systems. COS has defined additional categories of indicators such as loss of well control, loss of station keeping, mechanical lifting, and other process safety related indicators.

RBPS Element 19: Auditing

Auditing provides for a periodic review of process safety management system performance. The aim is to identify gaps in performance and identify improvement opportunities, and subsequently to track closure of these gaps to completion. Auditors may be second party (company personnel but independent of operational roles at the site) or third party (independent company) depending on company policies or local regulations.

Auditors must have the requisite skills for the upstream activities under audit. While all audits should address the effective implementation of the safety management system, there are technical aspects that also need to be addressed such as the process safety barriers status, and if offshore, key marine systems. COS (2014) has developed auditing requirements as well as requirements for the accreditation of Audit Service providers and auditors for SEMS audits, which have been incorporated by reference by BSEE.

RBPS Element 20: Management Review and Continuous Improvement

Management review and continual improvement is the practice where managers at all levels set process safety expectations and goals with their personnel and review performance and progress towards those goals. This may take place in a leadership team meeting or one-on-one with the relevant line manager.

Management review and continual improvement are the closing aspect in the so-called PDCA management loop of Plan-Do-Check-Act.

The review formalizes the link between goals set and results achieved. Where goals are not achieved then additional actions or investment can be included in the annual plan to correct the deficiency. Where goals are achieved, then new actions or investments can be agreed upon so that process safety performance is not static but improves over time. This is continual improvement.

3.3   CONCLUSION

The RBPS process safety management structure is an accepted industry practice. It provides a useful basis for upstream process safety or for enhancing upstream safety management systems via the appropriate inclusion of better tools, techniques, and systems from RBPS. It is non-prescriptive and can be incorporated into a company-specific management system compatible with regulatory requirements globally.

There are many other management systems that have varying numbers of elements. Some combine topics into fewer elements; others expand these out giving more elements. The number and arrangement are not important so long as the topics are covered. RBPS highlights some key topics not included in other systems.

As is noted in the pillar Learn from Experience, those interested in process safety should strive to learn from their own experience as well as the experience of others. All of the aspects of RBPS should be considered as an opportunity to learn and to improve process safety performance, whether or not they are regulated at your specific operating location.

4

Application of Process Safety to Wells

4.1   BACKGROUND

This chapter addresses the primary equipment, risks, and key process safety measures involved in drilling, completions, workovers and interventions, referred to collectively as well construction. It shows how the concepts contained in RBPS can be applied to further enhance current design, operational practices, and process safety performance. Special attention is given to well control and the safety barriers for preventing and mitigating a loss of well control event. The final section reviews selected process safety tools applicable to well construction.

Well construction terminology makes use of many unique terms and acronyms that may be unfamiliar to those new to the upstream industry. Basic definitions are included here to enable the reader to understand the topic, but for full definitions one may refer to industry reference books (see list below).

In this book, the terms drilling, completion, workover and intervention are derived from the Schlumberger online glossary (www.glossary.oilfield.slb.com). All the terms are associated with a potentially pressurized well that must be managed to prevent a loss of containment event.

Drilling: The process of creating a wellbore down to the reservoir production zone, including casing and cementing to achieve all needed isolations.

Completion: The collection of downhole tubulars and tools necessary for the safe and efficient production of hydrocarbons. It usually includes perforating the casing and isolating the well from the reservoir with explosives to provide access to the reservoir fluids.

Workover or Intervention: The repair or stimulation of a well to enhance its production rate. It usually involves inserting tools into the wellbore. A workover requires the presence of a rig, whereas an intervention does not.

Some process safety topics are common with production activities, covered in Chapters 5 and 6, and in some cases the reader is directed to those chapters rather than repeating the material in this chapter.

This section provides an introduction to the technology necessary to understand the application of RBPS to well construction activities. Section 4.2 then addresses risks and key process safety measures, while Section 4.3 covers process safety methods and tools.

Details on well construction technology may be obtained from industry references, including the following.

- *IADC Drilling Manual*, Vols 1 and 2 (IADC, 2015a)

- *Fundamentals of Drilling Engineering* (SPE, 2011)
- *Petroleum Engineering Handbook Vol 2: Drilling Engineering* (SPE, 2007)

Additional references from both IADC and SPE are available via their respective websites. ABB (2013) published a handbook that provides a simpler overview of the complete upstream production process, including details on the reservoir and well.

Multiple recommended practices (RP) related to well construction and well integrity are available from API for onshore and offshore wells (regularly updated and summarized in API, 2015), Norsk O&G (2016) for offshore well integrity, and NORSOK (2013) for well designs achieving two barriers. In addition to these standards and RPs, most large companies produce their own company specific well construction manuals. These may exceed standards and RP requirements based on company experience and industry good practices.

In some ways a loss of well control is similar to an abnormal situation for a downstream processing plant. The event must be recognized and addressed before the situation escalates to something more serious. Generally, more time is available to deal with a kick event than, for example, many runaway reaction situations, but the indication may be less obvious.

Some relevant incident descriptions are provided in this chapter to highlight examples and the possible application of RBPS. The first incident is the Deepwater Horizon loss of well control incident, named after the rig involved. It is also widely known as the Macondo incident, based on the prospect name. This book standardizes on Deepwater Horizon.

4.1.1 Drilling the Well: The Well Bore

The most important geologic, reservoir, and geomechanics factors related to process safety and loss of containment are pore pressure and fracture gradient, which are unique to each well. A summary of pore pressure and fracture gradient terms follows, including partial reference to the Schlumberger Oilfield Glossary.

Pore Pressure: The pressure of the subsurface formation fluids, commonly expressed as the density of fluid required in the wellbore to balance that pore pressure. In reservoir zones which have sufficient permeability to allow flow, this is the pressure of the hydrocarbons or other fluids trying to enter the wellbore. Safe well design balances the reservoir pressure with drilling muds of adequate density such that the mud hydrostatic pressure at the reservoir is sufficient to prevent inflow.

Fracture Gradient: The pressure required to induce fractures in rock at a given depth. If the fracture gradient is exceeded, then some of the dense drilling mud can be lost into the formation leading to a potential loss of hydrostatic head. If the pressure exerted by the hydrostatic head falls below the local pore pressure in the reservoir zone, then hydrocarbons can flow into the well.

Incident: Deepwater Horizon, April 2010

The Macondo well is located in the Gulf of Mexico. During actions for a temporary abandonment of the well, several failures occurred. The final cementing used a novel formulation, and this failed to seal the well. The heavy mud barrier was partially circulated out and an underbalanced situation resulted. Kick signals were misinterpreted, and a loss of well control followed. Flammable oil and gas were initially diverted into the mud room, but this soon ended up on the drill floor, where it ignited causing 11 fatalities, total loss of the drill rig, and the largest oil spill in US history.

Process Safety Issues: The Deepwater Horizon had an excellent occupational safety record. There was a program to address process safety (see later discussion on tools such as drill well on paper (Section 4.3.2)), but occupational safety was emphasized more than process safety. There were multiple technical defects identified relating to the cement job, the kick detection, and the apparent failure of the BOP. In fact, the BOP includes multiple safety systems, and some worked properly. The variable bore rams closed, sealed, and held considerable pressure. The shear ram failed to close because three drill string pieces were in the ram and the cutting face could not cut all the pipe in its bore (DNV, 2011). The National Commission made the following key conclusion.

> "The immediate causes of the Macondo well blowout can be traced to a series of identifiable mistakes made by [the companies involved] that reveal such systematic failures in risk management that they place in doubt the safety culture of the entire industry."

Source: Deepwater Horizon National Commission, 2011

RBPS Application

Process Safety Culture: Requiring a focus on process safety, not only occupational safety. Insufficient attention was given to a potential loss of well control with many other conflicting objectives present.

Asset Integrity and Reliability: Ensuring that safety critical equipment such as a BOP functions reliably is fundamental to process safety. While the BOP did function, it was presented with a condition which exceeded its design capability and it failed to seal the well.

Contractor Management: A main characteristic of well construction operations is the close relationship and dependency of the owner/operator, the drilling contractor, and the other specialty contractors. Establishing a clear understanding of who does what in routine, non-routine, and emergency situations is imperative.

As both are key factors in a safe design, these must be established first in the well design process. There are many sources used to determine these two factors including information from nearby wells already drilled, geologic studies, regional geologic studies, seismic studies, and others. In exploration wells, where there are no nearby wells, these factors are based on more indirect methods like seismic and regional studies. Drilling is a dynamic process and both factors are monitored to determine if initial estimates match actual conditions. This monitoring includes drilling penetration rates, drill cuttings characteristics, and well logs. (IADC, 2015a) A key part of drilling safety is the continual assessment of actual fracture gradient and pore pressure and other drilling parameters compared to the plan and adjusting the plan accordingly. Significant changes should be the subject of a formal Management of Change (MOC) review, a part of RBPS.

During drilling, the drilling team balances the need to prevent formation fluids from entering the well with the need to prevent excessive pressure from fracturing the formation. These counteracting requirements are achieved using drilling muds with adequate calculated density to achieve necessary pressures. A simplified version of the pore pressure / fracture gradient graph from the National Commission Chief Counsel's Report (2011) is presented in Figure 4-1. In this figure the pore pressure is the curve to the left and the fracture gradient is the curve to the right. More detailed versions of this figure show the casing used and the safety margin. The drilling team aims to maintain mud weight between the two curves for conventional overbalanced drilling. Casing and cement are used to achieve mechanical isolation between wellbore zones. Lighter drilling fluids can be used, and this is termed underbalanced or managed pressure drilling, but this is an advanced topic beyond the scope of this book. Casing protects the upper formation from fracture pressures that develop due to the heavier muds required to control pore pressure at lower depths.

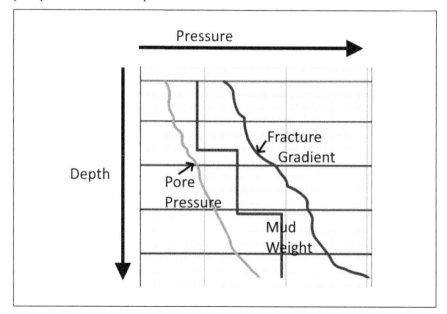

Figure 4-1. Pore pressure versus fracture gradient relationship

If formation fluids enter the wellbore during overbalanced drilling, this is called a kick. Hydrocarbons are the greatest risk, but saltwater is also undesirable. Hydrocarbon inflow can also occur during managed pressure or underbalanced drilling, and this is controlled using a managed pressure drilling system. A kick that is not recognized and managed can lead to a loss of well control event. Thus, the drill team must be continually observant for signs of a kick. Important factors contributing to kicks are swabbing and surge during trips in and out of the hole. Swabbing can result in an influx if the drill pipe is tripped out of the well too quickly as this reduces the hydrostatic head below the local pore pressure. An opposite issue is surge. This can occur if pipe is moved too quickly into the well creating a surge pressure that exceeds the fracture gradient, resulting in a loss of mud to the formation, a reduced mud weight barrier, and subsequent influx. There are other causes of kicks that are beyond the scope of this concept book. The reader can refer to the IADC and SPE drilling texts listed earlier in Section 4.1 for additional discussion of kicks.

4.1.2   Drilling the Well: Barriers

Well integrity is defined at its simplest (Norwegian Oil & Gas Assn, 2012) as a condition of a well in operation that has full functionality and two qualified well barrier envelopes. Any deviation from this state is a well integrity issue. What constitutes a barrier envelope is a complex topic and differs for different well activities even on the same well. A full discussion on the term 'barrier' and some differences between standards and onshore vs offshore usage was provided in Chapter 2.

A document defining in diagrammatic form how to achieve two qualified barriers for many different well activities is given by NORSOK D-010 (2013). One example is shown in Figure 4-2, which covers drilling, coring and tripping activities with a shearable string. A shearable string means that the drill string can be cut and isolated using the BOP blind shear ram.

The figure shows there are two main barriers preventing a loss of well control during drilling – the fluid column being the primary one and internal well components (the formation, casing, set cement, wellhead, high pressure riser and the BOP) being the secondary. The figure lists but does not show the riser between the seabed and the surface due to issues of size. The riser is a complex flexible connector carrying well fluids upwards, but also mud, instrumentation and other utilities to the well. An onshore well has very similar barriers / barrier elements – but no riser. Additional details in NORSOK D-010 show how to qualify / verify each barrier.

As loss of well control is the most critical process safety risk, the key barriers are discussed in some detail in this chapter. These barriers also protect against other hazards described in following sections.

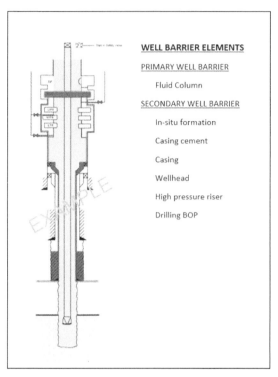

Figure 4-2. Two-barrier diagram for drilling, coring and tripping with a shearable string

4.1.3 Drilling the Well: Fluid Column

The fluid column with sufficient hydrostatic pressure is one complete barrier on its own. It is a mixture of fluids (water-based, non-water based, or gaseous) and solids engineered to specific densities, collectively called "mud", to match the requirements of the pore pressure and fracture gradient curves (Figure 4-1). Lower depths in the wellbore require higher density drilling mud and casing sections isolate higher portions of the well where the mud pressure exceeds the fracture gradient. In offshore US federal waters, BSEE generally requires a drilling margin of 0.5 pound per gallon (i.e., 0.5 ppg) below the lowest estimated fracture gradient to provide a safety margin. As previously mentioned, drilling should not be thought of as a static situation; conditions change requiring response to maintain the correct mud weight.

Mud returns from the well carrying drill cuttings have the soil/rock cuttings separated using shakers to allow cleaning, reconditioning, and reuse of the mud. Careful monitoring of mud flowrate is necessary to determine if there is mud loss into the formation or an influx of reservoir fluids into the wellbore.

4.1.4 Drilling the Well: Casing

Casing is normally composed of sections of steel pipe screwed together. Usually multiple diameters of casing are employed in decreasing size with well depth to allow new casing sections to pass through already set sections. Typical casing naming conventions in order of depth are: 1) conductor, 2) surface, 3) intermediate, 4) production, and 5) reservoir liner. Not all wells have all these types of casing; some wells have drilling liners at intermediate depths in addition to the reservoir liner and some wells have the production casing thru the reservoir and no liner.

The conductor casing protects the well from collapse from loose near-surface aggregates and serves as the foundation for the well. For onshore wells the conductor is often 15-30 m (50-100 ft) deep, for offshore wells it may be 300 m (1000 ft) deep. Surface casing comes next, and it protects local groundwater resources from potential contamination from well fluids and typically extends at least 15 m (50 ft) below any potable groundwater unless local regulations require more. Surface casing is usually cemented all the way back to the surface or seabed (i.e., a layer of cement outside the casing separating it from the formation), completely isolating any groundwater resource. Testing of cement integrity is required once completed.

In some cases, the total well depth is safely drilled from the surface casing alone, but usually another deeper intermediate casing is required. Intermediate casing protects the wellbore from multiple problems and ensures that the pore pressure fracture gradient limitations are not violated. The production casing is usually run from the wellhead to the full design depth of the well.

An additional process safety hazard is due to the mechanical handling of casing or drill pipe segments near to producing facilities. Dropped segments can rupture pipework or vessels and create a LOPC event. This is a SIMOPS issue and special controls are needed, which are discussed further in Section 5.2.5.

4.1.5 Drilling the Well: Cement

Cementing and cement compositions are discussed in Chapter 9 of the SPE Petroleum Engineering Handbook (2007).

Cement is used to permanently seal annular spaces between the casing and the borehole walls. Cement is also used to seal formations to prevent loss of drilling fluid and for operations ranging from setting kick-off plugs to plug and abandonment. Various additives are used to control density, setting time, strength, and flow properties. The cement slurry, commonly formed by mixing cement, water and assorted dry and liquid additives, is pumped into place and allowed to solidify (typically for 12 to 24 hours) before additional drilling activity resumes.

4.1.6 Drilling the Well: The BOP

The blowout preventer, BOP, is a safety device that forms part of the well barrier system (see Figure 4-2). The terms BOP, blowout preventer, blowout preventer stack and blowout preventer system are used interchangeably. Note the BOP normally requires manual actuation, except in the case of a drift off event offshore, and thus it is a combined hardware and human barrier. There are multiple designs for a BOP. The BOP includes devices protecting both the drill string and the annular space, and it must allow for passage of all drilling equipment and tool sizes necessary to construct the well.

The BOP has a number of operational uses as well as its process safety function which is preventing uncontrolled releases from a well and controlling bottom hole pressure while killing a well. In bow tie terms, the BOP appears as either a prevention barrier or as a mitigation barrier depending on which function or when the BOP is actuated.

The BOP protects against well fluid flow through the downhole tubing string or through the annular space. IADC (2015a) provides additional detail on BOP systems for offshore applications. Well fluid flow outside and around the well casing system relies on the formation strength, casing and cement barrier elements as shown in Figure 4-2. In offshore US federal waters, BSEE has detailed requirements for BOPs as outlined in the Final Well Control Rule (30 CFR 250). Other regulators have their own requirements. In the US, onshore wells are not regulated by BSEE and these wells follow API recommendations and any local regulations.

Offshore BOPs are normally described as a stack because several devices are located atop one another (Figure 4-3). The BOP stack with its associated connections can be quite tall – the dimensions of the main elements add to 36 ft (11 m) in this example. Onshore BOPs are usually more compact. The topmost stack in an offshore BOP is the annular ram which closes around the drill string acting as a lower pressure flow containment device. It protects against smaller scale kicks moving up through the annular space. In one common annular ram design, a donut shaped elastomer packer is forced around anything in the hole. During drilling operations, the annular is normally fully open to allow return of mud fluid with drill cuttings. Drill pipe and tool joints can be moved through the annular ram while it is closed.

Beneath the annular preventer are the ram preventers. IADC (2015a) notes there are five main types of rams: blind rams, pipe rams, variable bore rams, shearing blind rams, and casing shear rams. The number of rams installed depends on several factors and can be up to 8 in number. Blind rams are flat and seal the wellbore when no pipe is present. Pipe rams are curved and seal the annular space around specific drill string or casing elements. Variable bore rams perform a similar function but seal around multiple diameters. Blind shear rams are designed to cut the drill string and create a seal. Casing shear rams cut casing elements but are not designed to seal.

In the past, a single blind shear ram was used. If a tool joint (i.e., the connection between two drill pipes) was located in the ram, it would be necessary to lift the drill string to move the tool joint out of the cutting zone as the shear ram is designed to cut the drill pipe but not this thicker joint. This action could be impractical in an emergency situation. BSEE now requires the use of dual blind shear rams so that shearing can occur even if a tool joint blocks one shear ram from functioning as the other will only have drill pipe inside. Onshore practice in the US varies but usually employs fewer rams than offshore.

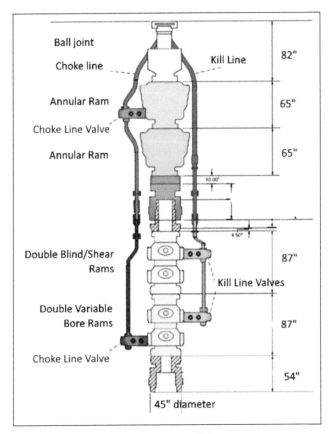

**Figure 4-3. Example offshore BOP
(derived from Shell, 2015)**

Further elements of the BOP system are the choke and kill lines. These pipe connections are used to terminate a blowout event but may also be used for measurements to determine if an influx is occurring. API 16C (2015c) provides specifications for choke and kill equipment.

Subsea BOPs include electro-hydraulic pods which are a combination of computer and hydraulic control valves. The rig sends control signals to the computer which activates the subsea hydraulic valves and uses the hydraulic energy stored subsea to activate the rams in the BOP. There are two automatic systems legally required in the US: one for a disconnect of the riser and one for loss of signal from the rig. Other than these two events, BOP actuation is manual. Subsea BOPs are controlled with dual redundant pods – often termed blue and yellow. These respond to signals from the driller on the surface and direct hydraulic fluid to operate the multiple BOP systems. Remotely Operated Vehicles (ROVs) are used to address any problems with subsea BOPs and to enable functioning of the BOP (by manually turning valves) when built-in redundant systems have failed.

4.1.7 Well Completions

Well completion equipment is designed to withstand all expected loads throughout the well life. During completion activities, the wellbore is filled with a dense completion fluid to maintain sufficient hydrostatic pressure. Many completion sequences provide for removing the heavy weight mud so that the well is perforated through tubing in a clean fluid.

The BOP is removed and replaced with a wellhead valve assembly termed a Christmas tree. The wellhead includes a surface safety valve that terminates flow if required. Offshore wells often install a sub-surface safety valve (SSSV), typically 30 m (100 ft) below the seabed, a regulatory requirement in the US. Onshore sour gas wells may also employ SSSVs. SSSVs are not considered a permanent barrier. They may be used as a temporary barrier in combination with a mechanical barrier (e.g., plug or back-pressure valve) for BOP or Christmas tree removal.

The annular space between the production tubing and the casing is sealed, normally using a mechanical sealing device to prevent hydrocarbons from flowing up the annulus.

4.1.8 Well Workovers or Interventions

Workovers and interventions are similar activities with the aim to maintain, restore or increase productivity of a well or ensure well integrity. A workover usually requires a drilling or completion rig with all its equipment to be placed over the wellbore, whereas an intervention uses equipment that may not require a rig. Both activities require tools of various types to be inserted into the well to do the required work.

A loss of well control is possible during well workovers or interventions. Dense fluid is used as the primary safety barrier during workovers. If the Christmas tree is removed, a BOP is installed. This provides the two safety barriers required. Alternative arrangements are employed if coiled tubing is used in a workover.

Explosive charges are used in some workover activities for cutting or creating entry points for hydrocarbon production or removing obstructions within the well and formal explosives safety practices are required.

4.1.9 Depleted Wells

When reservoirs have insufficient hydrocarbon reserves or have been depleted by production, the wells are plugged and abandoned. Wells that are to be left and returned to for later production are temporarily abandoned. In a temporary abandonment, a barrier is not required to extend across the full section of the well and include all annuli.

Permanent abandonment has several objectives: 1) provide isolation between hydrocarbon zones, 2) protect freshwater aquifers, 3) prevent migration of formation fluids through the wellbore, and 4) remove all surface equipment and, for offshore wells, cut pipe to below seabed. A process safety event occurs when any of these objectives is not met.

Plugging is normally achieved by multiple barriers. A dense abandonment fluid is pumped into each isolated zone with sufficient hydrostatic head to exceed any formation pressure. Cement plugs are set at the bottom of the well isolating any perforations. Higher plugs isolate higher sections of the well. These plugs can be supplemented with a mechanical plug. A surface cement plug is also set. Each region will have plugging and abandonment requirements.

4.2 WELL CONSTUCTION: RISKS AND KEY PROCESS SAFETY MEASURES

4.2.1 Overview

The major process safety hazard associated with drilling, completion, workover, and interventions is a loss of well control. This can result in a loss of containment event where subsurface hydrocarbons have the potential to escape uncontrolled into the atmosphere, land, waterways, ocean, or sub-surface strata. Consequences can include fire, explosion, toxic gas exposure, pollution, and aquifer contamination affecting people, the environment, assets, and company reputation. The primary cause for a loss of well control is failure of or lack of adequate barriers and/or lack of well control management and well data monitoring. Formation fluids can flow into the wellbore if the hydrostatic mud pressure is insufficient or is compromised (e.g., due to a loss of mud into the formation) to below the pore pressure. While kick events are a part of well construction, these are normally managed by closing the BOP and circulating out with higher density mud or another response. However, if not recognized, kicks can lead to a loss of well control. Other causes of loss of well control include, but are not limited to, intercepting an existing well with new drilling; casing or drill string separation; corrosion or mechanical erosion; drilling into a higher pressure zone; earthquakes; fault movement; and premature detonation of shaped charges.

Other hazards associated with well construction include loss of containment events from improper operation or well equipment failures at the surface such as in mud separation rooms, separation and treatment facilities, and pumping and compression. This can result in the accumulation and potential ignition of flammable gas concentrations or liquid pools in equipment spaces such as the drill cabins, the control rooms, and other occupied spaces. Depending on the well location, offsite persons can also be impacted.

Loss of control of energy associated with well construction is also a serious issue, but the outcome is more often a personal injury event rather than a process safety event, and so is not covered here. Examples include loss of control of heavy rotating equipment, dropping of drill string or casing segments, or unravelling of stored energy in coiled tubing.

The following sections summarize risks and highlight key process safety measures during well construction. The key process safety measures listed provide a selection of the more important elements to illustrate the application of RBPS to well construction. Other RBPS elements may also apply to some degree.

4.2.2 The Well

Risks

The loss of well control is the primary risk related to well construction. The geologic and geomechanics factors and well construction technologies are very diverse. This concept book focuses only on those that can cause loss of well control. IADC (2015a) in its chapter on drilling practices reviews many potential drilling dysfunctions and provides suggested remedies not covered here.

SINTEF, a research group in Norway, maintains a subscription database of blowout events. A summary is provided online covering the period 1980-2014 (https://www.sintef.no/en/projects/sintef-offshore-blowout-database/). This shows that there were 292 blowout events recorded in that period in the UK / Norway and in the US GOM. A breakdown by activity shows 56% occurred during drilling, 7% in completions, 23% in workovers, 11% in production (mostly due to external event), 5% in wire-lining, and 1% in abandonment. Thus, all phases of well operations are subject to potential blowout.

Fluid Column: The loss of mud hydrostatic pressure in the fluid column can lead to a kick. This can occur if the drill pipe is removed too quickly (swabbing) or if fracture gradient of the formation is exceeded as discussed in Section 4.1.1. Loss of mud hydrostatic pressure can also occur for offshore wells if the riser (connecting the surface drilling facility to the subsea BOPs at the seabed) is ruptured as might occur due to a collision with a support vessel. The riser provides, amongst other functions, the annular space for mud to return to the surface mud room. A rupture allows mud to flow out into the sea causing a loss of hydrostatic head. Additionally, loss of hydrostatic head can occur from connection failures in the riser, failures at the lower marine riser package (LMRP), or failures in the well casing connections or segments which can occur from drilling wear or other factors. When using synthetic oil-based muds, small gas influxes may dissolve in the fluid which are not as easily detected as a kick. Loss of well control can occur due to poor planning resulting in an incorrect mud density combined with a lack of well condition monitoring to detect an influx.

The API 65-2 standard (API, 2010) describes several examples of kick events. The most common cause of kicks in the US historically is swabbing – i.e., while lifting of the drill string (see Glossary). This can create an effect like a piston withdrawal causing a zone of low pressure below, allowing formation fluids to enter the wellbore.

Casing: API RP 100-1 (API, 2015) for onshore hydraulic fracturing notes that casing design and selection is critical to well integrity including well control. It must be designed to withstand all anticipated loads while running into the hole, as well as loads during drilling, completions, workovers, interventions and production.

The prime design factors on casing are ratings for tension, burst and collapse pressure. The selection of casing material is important to avoid corrosion and loss of containment events. IADC (2015a) and API 5CT (2019b) provide guidance on material selection to deal with sour gas, CO_2, chlorides, temperature, carbonate concentration, and produced water contaminants.

Cement: Cement is a critical barrier element in achieving isolation and multiple local factors can affect cement integrity. The failure of the cement job to achieve the required isolation in temporary abandonment of the Deepwater Horizon rig was a significant contributing factor to the blowout.

Dusseault et al (2000) discuss mechanisms causing onshore oil wells to leak, especially those related to cement failures. They identify cement shrinkage as an important factor, and this leads to channeling and high cement permeability. Inadequate design or installation and contamination are all important factors in cement failures.

The BOP: While multiple responses are possible to a kick event, one common response is to circulate mud using the Driller's Method prior to use of the BOP. This requires two complete separate circulations of drilling fluid in the well. The first circulation removes influx with original mud weight, while the second uses kill weight mud. If this is not successful, the well is sealed using the BOP. A simpler method, the Engineer's Method (aka 'Wait and Weight'), requires only one circulation of a heavier mud weight material.

As previously noted, a BOP normally requires manual actuation, except in the case of a drive off or drift off event offshore or a loss of control signal. To be effective in stopping a blowout, the BOP must be actuated in a timely manner. A blowout preventer will not stop the blowout if it is not actuated in a timely manner. This was apparent in the Deepwater Horizon event, where the drill string was pushed off-center inside the BOP during the event and could only be squeezed but not cut by the blind shear ram (DNV, 2011).

If the BOP fails to seal and well fluids rise to the surface, then a diverter valve can be actuated directing well fluid flow overboard (offshore) or to a flare or burn pit (onshore) away from the locations where crews are working. This reduces the risk of harm to people but does not eliminate the hazardous situation entirely.

Key Process Safety Measure(s)

Process Safety Competency: Well construction is dynamic and manually controlled. The competence of those involved to be able to conduct operations and detect and respond promptly when an unplanned influx occurs is key to safe well construction operations.

Incident: Pryor Trust Well Blowout, Oklahoma, Jan 2018

A blowout and rig fire occurred at a Pryor Trust gas well in Oklahoma. The fire killed five workers, who were inside the driller's cabin on the rig floor. The blowout occurred about three-and-a-half hours after removing drill pipe ("tripping") out of the well.

Process Safety Issues: The drillers were operating in an underbalanced condition and a dense slug intended to increase the mud density was miscalculated leaving the well still underbalanced. This created a condition allowing formation fluids to enter the wellbore. The alarm system was completely disabled, and indications of a kick were missed. The drilling contractor did not follow the company-required flow check procedures. The company's onshore rigs are not covered by offshore SEMS requirements and the safety management system was not effective.

Soon after the blowout commenced, it ignited, and the hydraulic lines used to actuate the BOP were destroyed by the fire, preventing the drill team from being able to shut-in the well. It is important that process safety devices can survive the event they are designed to protect against.

Source: CSB, 2019

RBPS Application

Hazard Identification and Risk Analysis: Knowledge of hazardous events and their potential consequences is important to assess the adequacy of safety devices, including the security of utilities enabling the device to function.

Training and Performance Assurance: Loss of well control events are preceded by kick events that must be identified by the drill team and appropriate responses implemented to prevent a loss of well control and the potential for a blowout.

Learn from Experience: There are many similarities between the Pryor Trust event and both the earlier Deepwater Horizon and Montara events. If those lessons had been learned sufficiently, then the Pryor Trust event could have been prevented.

Conduct of Operations: This RBPS element is about structuring tasks, ensuring they are performed correctly, and minimizing variations in performance. This element is important to well construction operations because they are manually controlled, involve numerous people/organizations, and are carefully balanced between control and loss of control.

4.2.3 Shallow Gas

Risks

Shallow gas deposits near to the surface can be encountered during drilling before the BOP and surface casing are in place and can lead to a shallow gas well control incident. This can be associated with either onshore or offshore drilling. Oil is not normally present with shallow gas. While mud weight can be increased, if this fails then it may be necessary to drill a relief well to kill the shallow gas flow.

During a shallow gas incident, it is not normally advised to try to shut in the well as the surface formation is not strong enough to provide for containment. A safer option is achieved using a diverter valve to direct flow away from the rig floor.

Key Process Safety Measure(s)

Hazard Identification and Risk Analysis: Identification of shallow gas is key to understanding and managing the risks. Shallow gas is hard to detect, but newer digitally enhanced seismic analysis can reveal this hazard. Consequences onshore are primarily related to flammable and potentially toxic gases. Offshore consequences are similar as shallow gas can bubble to the surface under a floating drilling rig and create a flammable atmosphere. It can also damage the sea floor and destabilize a jack-up rig. A shallow gas incident can damage or rupture the drill string and thus reduce the ability to deliver a heavier mud to the problem zone. Personnel require evacuation which can be difficult due to the flammable atmosphere but, as seen in the Snorre A blowout described in the following incident description, can be done successfully.

Shallow water hazards are similar but without the flammable or sour gas hazards. They are thus more of an operational than process safety problem. However, depending on the source of the water and if there are nearby receptors, there can be a pollution risk. For example, if the water source is due to accumulation of nearby water injection wells, then the water may be contaminated.

4.2.4 High Pressure High Temperature (HPHT) Wells

Risks

Loss of well control risks are heightened when well construction involves high pressure, high temperature (HPHT) reservoirs. These have temperatures exceeding $300°F$, a pore pressure of at least 0.8 psi/ft, or requiring pressure control equipment exceeding 10,000 psi. Drilling, completions, workovers, interventions and abandoning wells in HPHT environments are at greater risk due to the complexity associated with the high pressure and high temperature and having a higher probability of well control incidents and equipment failures.

Incident: Snorre A Blowout, Norway North Sea, November 2004

Snorre A was a large integrated tension leg platform with processing, drilling and accommodation modules. Activity levels were high with SIMOPS covering production, drilling and well intervention underway. During a workover operation prior to further drilling in a well, a gas blowout occurred on the seabed with the subsequent gas flowing to the surface and under the facility. Ignition did not occur. There were 216 persons on board at the time of the incident, 181 of whom were evacuated to other installations while the other 35 persons remained on Snorre A to carry out emergency response and well control tasks. The gas blowout was stopped and the well brought under control the day after. No one was injured in connection with the incident.

Process Safety Issues: Complex defects in the well due to corrosion and other factors were not sufficiently managed. Shore-based HAZOPs to address the problems being encountered were carried out but not communicated to offshore personnel. The Petroleum Safety Authority (PSA) identified multiple safety barriers that failed, and these allowed the incident to occur. The PSA concluded that total loss of the facility was possible and that this serious near miss was one of the worst events in Norway.

Source: PSA, 2005

RBPS Application

Process Safety Culture: Multiple organizational issues were identified including too slow integration of Snorre into the Statoil organization following the acquisition of Saga Petroleum, critical questioning of operations was not welcomed, and management was not sufficiently engaged. RBPS suggests how to enhance process safety culture.

Asset Integrity and Reliability: Offshore personnel allowed the BOP to be partly disabled as only the annular preventer was available. RBPS offers guidance on means to ensure full availability of critical barriers.

Key Process Safety Measure(s)

Conduct of Operations: Significant planning and engineering is required to work in an HPHT environment, including the specification and use of equipment and drilling and completion fluid and cement specification. Personnel should be trained in HPHT operations.

4.2.5 Adjacent Wells

Risks

A serious threat both onshore and offshore is the potential to intercept or collide with other existing wells, which can result in a loss of well control event. Limited well spacing at the surface increases the likelihood of a possible collision with other wells.

Key Process Safety Measure(s)

Process Knowledge Management: To avoid this hazard, good mapping of all wells is necessary, and a drill plan developed that maintains adequate separation.

Operating Procedures: Where collision is considered a high risk, good practice is to shut in potentially affected wells until the possibility of collision risk is low or eliminated.

4.2.6 Completions

Risks

Loss of well control is possible during the completion phase of well construction. This was discussed in Section 4.2.2 where SINTEF data shows 7% of blowouts occur during completions. Understanding the risk is crucial as completions normally consist of changing out fluids in the well that maintain the hydrostatic pressure and use of different equipment than during the drilling phase that serve as the second barrier in preventing a loss of well control. An additional risk relates to shaped charges used to perforate production casing to allow formation fluids to flow into the well. Perforation is an activity that has process safety as well as personnel hazards.

Key Process Safety Measure(s)

Safe Work Practices: The avoidance of loss of well control through the use of two or more barriers is a critical requirement, similar to other phases of well construction. Proper handling and grounding systems to prevent premature detonation is also required. Transportation, storage, loading and unloading perforating guns also require attention per IADC guidance.

4.2.7 Workovers or Interventions

Risks

Blowout risks are important during workovers and interventions. The SINTEF data shows 28% of blowouts occur during these activities. Swabbing, as noted before, is a most common cause of kicks and careful control of fluid flowrates and tubing string or tool removal rate during workovers and interventions is important.

A significant risk in workovers is that the producing open formation can be depleted and can be close to other producing formations that are not depleted. These wide pressure variations in close proximity can result in losing fluid and taking a kick from the same completion zone.

Workovers and intervention activities may also use explosive charges to rejuvenate a well, similar to completions. Improperly grounded systems can result in premature detonation of the explosive charges with damage to the wellbore systems or to personnel if accidentally triggered at the surface.

Key Process Safety Measure(s)

Safe Work Practices – same as Section 4.2.6

4.2.8 Depleted Wells

Risks

The SINTEF data reported in Section 4.2.2 shows 1% of blowout events occur during abandonment. Plugging wells in depleted reservoirs can involve multiple barriers. These barriers can fail allowing hydrocarbons to flow and eventually to exit the well or into the surrounding soil. Dusseault et al (2000) discuss reasons for loss of integrity and sustained casing pressure in oil wells.

Key Process Safety Measure(s)

Asset Integrity and Reliability: As with all equipment, barriers must be properly designed and maintained to ensure their long-term performance, with critical systems also providing a means to warn operators of impending or actual failure.

4.2.9 Mud Room / Shale Shakers, Well Fluid Handling and Treatment Locations

Risks

Release of flammable and sometimes toxic gas from the well mud returns or treatment systems is possible. Gas clouds can result where there is insufficient ventilation and fires and explosions can result if ignition occurs. Large scale onshore facilities which are fully enclosed have well developed ventilation systems for mud handling and treating systems. Onshore drilling and production facilities are usually well separated. Offshore, some facilities conduct both well construction and production simultaneously and a SIMOPS plan is needed to address potential risks (see Section 5.2.5 for further details).

Key Process Safety Measure(s)

Hazard Identification and Risk Analysis: Companies with offshore deepwater or large onshore production facilities often carry out detailed fire and gas safety studies. This may be less common at smaller onshore facilities or unmanned shallow water installations. The HIRA studies apply to all aspects of well construction, not just surface handling. Provision of flammable and, if appropriate, sour gas detection is standard, with alarms to the drill cabin as well as to the control room.

For onshore facilities where there is simultaneous drilling and production, it is common to carry out facility siting studies. This is particularly the case if the inventory of hydrocarbons exceeds the nominated threshold limit and OSHA PSM (1910-119) applies. Facility siting studies are not common for temporary well construction facilities with no associated permanent production. Refer to Section 5.3.2 where the topic of facility siting is discussed in detail.

4.2.10 Surface Process Equipment at Well Construction Facilities

Risks

There are multiple potential sources of flammable gas or liquid releases from surface kick handling and mud process equipment at well construction sites when the drilling reaches hydrocarbon zones. These are mud degassers, solids control equipment, and mud tanks and are common at all drill sites. There also can be gas and other hydrocarbon handling equipment at drill sites that conduct well flow backs or underbalance drilling.

Liquids storage, if any, is usually in atmospheric storage tanks that store hydrocarbon liquids (generally mixtures of C5 and heavier). This material is normally transported by truck, especially for remote areas as well as exploration wells that do not have the export infrastructure of production facilities. However, this is generally associated with limited well flowbacks or underbalance drilling which are not common in most drilling outside of shale developments. Wells undergoing workovers or interventions are usually close to production facilities and thus have SIMOPS risks. Offshore FPSO and FLNG facilities also store hydrocarbon liquids. Storage risks are discussed in Chapters 5 and 6.

Key Process Safety Measure(s)

Compliance with Standards: Process equipment should be designed in accordance with industry standards and company practices with the intent to provide integrity and minimize release potential.

Hazard Identification and Risk Analysis: Potential leak scenarios should be identified, and safeguards reviewed for adequacy, with potential hazard zones estimated to establish risks to personnel, other process equipment, or to the affected public. CCPS (1999) provides suggested means for how to predict hazard zones.

4.2.11 Harsh Weather

Risks

Harsh weather, both such as tornadoes for onshore and storm winds and waves for offshore, can lead to loss of containment events. Some harsh weather offshore events which a reader might wish to examine include the following (multiple references on-line).

- Alexander Kielland, Norway 1980, 123 fatalities
- Ocean Ranger, Canada 1982, 84 fatalities

- Kolskaya jackup rig, Russia 2011, 53 fatalities

Offshore, harsh weather events might result in rupture of the drilling riser due to drill vessel movement or collision with a service vessel. Rupture of seabed pipework can be caused by subsea or loop currents, seabed movement, or collisions with dragged objects from facilities that have lost station. Production riser failures and subsea infrastructure vulnerabilities are more critical during the production phase rather than during well construction. This discussion is provided in Chapter 6.

Onshore, harsh weather can also cause hazards to onshore facilities, such as toppling of poorly anchored land rigs.

Key Process Safety Measure(s)

Safe Work Practices: Many companies establish a weather window both onshore and offshore to limit operations such as use of cranes, people transfer by boat, and helicopter operations that might be adversely affected by strong winds, high seas, etc.

Emergency Management: It is common in areas subject to hurricanes or typhoons for MODUs to be moved before a storm strikes, and this reduces personnel risks due to structural failure or sinking. The North Sea, Canada, and Alaska all have long periods of harsh weather, but not with as strong winds as during a hurricane, and evacuation or rig relocation is not generally required.

4.2.12 SIMOPS

Risks

Ineffective management of Simultaneous Operations (SIMOPS) can introduce new and significant risks versus each operation on its own. SIMOPS occurs when two or more separate operations are occurring that can interact in potentially unexpected ways and create a hazard. Examples are drilling a well while producing from other wells in close proximity at the same time and conducting hot work on a production facility while drilling into a hydrocarbon bearing zone. Refer to Section 5.2.5 for more details on the topic of SIMOPS.

Key Process Safety Measure(s)

See Section 5.2.5 for a listing of process safety measures.

4.3   APPLYING PROCESS SAFETY METHODS IN WELL CONSTRUCTION

Various process safety methods and tools are applicable to upstream operations. This section outlines a few of the more important tools relevant to managing the risks of well construction.

To avoid repetition, the methods for well construction, onshore production, offshore production, and the design activity (Chapters 4, 5, 6, and 7, respectively) are presented only once. Topics such as fire and gas detection and fire protection equipment, which also apply to design and production activities, are covered in Chapters 5, 6 and 7. Approaches to onshore facility siting are discussed in Chapter 5 and the potential for flammable clouds to progress from flash fires to vapor cloud explosions is discussed in Chapter 6.

4.3.1 Regulations and Standards

Regulations

An overview of regulations was provided in Section 2.8. Specific to drilling for offshore well construction in the US, BSEE issues regulations under 30 CFR 250. These have extensive requirements as discussed earlier for the Final Well Control Rule. This Rule refers to several API standards including API 53 (2018) which has important guidance for BOPs. Before an offshore well, bypass, or sidetrack is drilled, the operator must submit detailed information to BSEE showing compliance with safety and environmental requirements. After review and acceptance, a permit to drill is issued.

The situation onshore in the US is less clear. As previously noted, BSEE regulations do not apply to onshore wells, and OSHA PSM and EPA RMP regulations do not apply to drilling. This gap in regulatory coverage is discussed at length by the CSB (2019) in their report on the Pryor Trust gas blowout. Companies strive for safe operations by following API standards, state regulations, and their own company procedures.

Internationally where goal-based regulations are used, the operator develops a safety case describing the risks and how the company plans to achieve a safe operation. For well construction, this may take the form of a well construction safety plan. Safety cases apply onshore and offshore. Some international oil and drilling companies follow their own internal safety case approach voluntarily as they believe a consistent global approach for their total operations is easier to manage and achieves higher levels of process safety than following only diverse local regulations. The BSEE SEMS Regulation is primarily goal-based as the actual management system is left to the company to design. It must meet the requirements of the regulation, which incorporates by reference API RP 75 and is audited following COS (2014) Audit Requirements guidance.

In consultation with international regulators who use goal-based regulations, IADC has developed an HSE Case approach specifically for well construction. This is to enable mobile offshore rigs to operate in different countries with minimal additional safety documentation required to obtain necessary permits. There are two versions of the HSE Case Guidelines (IADC, 2009, 2015b), one for onshore and one for offshore. They are broadly similar documents, but with appropriate differences to reflect their application areas.

Mobile Offshore Drilling Units (MODU) must follow three sets of regulations: flag state, coastal state, and class society rules (e.g., ABS, DNV GL, Lloyd's Register, etc.). The flag state is the country where the MODU is registered and the coastal state is where the MODU is currently operating. As MODUs are vessels, they must also comply with the ISM safety management code issued by the International Maritime Organization and adopted globally. This is certified for compliance by a class society. Class Rules apply primarily to design requirements and for periodic surveys to assure no degradation is occurring. The Regulator addresses, amongst other matters, drilling activities and process safety.

Codes and Standards

Well Construction is addressed in a large number of codes, standards and recommended practices, collectively called 'standards' here. API in a summary document (API, 2015a) lists over 100 standards related to onshore drilling with a similar number for offshore drilling. Many API standards are reissued as ISO standards so readers should not assume these are necessarily different. In addition to API standards there are standards issued by other bodies (e.g., IADC, ISO, IEC, Norway NORSOK, Australian Standards, Canadian Standards, etc.); however, API has the largest collection. The process for standards creation and approval is governed by an ANSI standard. In this book, examples are taken primarily from API, IADC and NORSOK.

Most API standards topics are technical. An exception is API RP 75 (2019a) which addresses a safety and environmental management system. API RP 75 is adopted into the BSEE SEMS II regulation making it a legal requirement for the US OCS as mentioned in Section 2.8. The RBPS system is covered in Chapter 3. One can adopt RBPS as a basis but is still required to comply with SEMS.

API 100-1 for onshore drilling addresses barriers but does not specify a target number. BSEE (2016) in its final drilling rule does specify a two-barrier requirement when removing the Christmas tree or well control equipment. The global oil and gas industry operating in Norway developed guidance (Norsk O&G, 2016) that is more explicit in specifying a two-barrier requirement for drilling.

For example, Figure 4-2 shows two barriers for drilling offshore. The first, the fluid column, is a complete barrier on its own, whereas the second is composed of six elements: the BOP, the casing, cement, the formation strength, high pressure riser, and wellhead. This is because the BOP and wellhead address loss of containment events coming up the tubular and the annulus, but not releases outside the casing. Thus all 6 elements are required for the second barrier to be successful. The situation onshore would be the same except no riser element is present.

Important barriers for loss of well control in well construction were covered in Section 4.2.2.

The two-barrier approach requires that barriers should be designed to enable rapid restoration if a barrier is lost; and if it is lost, then a return to two barrier functionality should be the top priority over other well construction operations. To achieve this, barriers must be explicitly defined, and their status known at all times, including an ability to test barriers. In practice, it may be difficult to know full barrier status at all times, especially in deep water, so regular testing may be necessary instead. For example, it is not possible to test a blind shear ram when there is drill string in the wellbore as this would seriously damage well equipment, but it can be tested when the drill string is removed. BSEE (2016) and API 53 define test and test frequency requirements for the BOP.

Company Practices

Most companies involved in well construction have their own practices or well operations manuals that are followed, except where more stringent local regulations apply. These manuals are developed based on a company's own well construction experience and aim to prevent or mitigate high consequence incidents.

In order for company practices to be effectively used, and the lessons learned from experience incorporated into new operations, both the operating company personnel and its various drilling, cementing, and other contractors need to be aware of the content. Where a company developing the reservoir and its well construction contractor both have practices, the practices should be aligned, or bridged, so that both are satisfied with the content and personnel involved know if there are any changes to their accustomed procedures. Both IADC and API have developed guidance on bridging documents that ensure the necessary alignment.

Guidelines

There are many process safety methods and tools documented in CCPS texts that are relevant to drilling and associated surface operations.

- *Guidelines for Hazard Evaluation Procedures* (CCPS, 2008a)
- *Guidelines for Asset Integrity Management* (CCPS, 2018a)
- *Guidelines for Technical Planning for On-Site Emergencies* (CCPS, 1995)
- *Guidelines for Evaluating Process Plant Buildings for External Explosions, Fires and Toxic Releases* (CCPS, 2012)
- *Guidelines for Preventing Human Error in Process* Safety (CCPS, 2004)
- *Human Factors Methods for Improving Performance in the Process Industries* (CCPS, 2007b)
- *Guidelines for Inherently Safer Chemical Processes: A Life Cycle Approach* (CCPS, 2019a)
- *Bow Ties in Risk Management* (CCPS 2018c)
- *Layers of Protection Analysis: Simplified Process Risk Assessment* (CCPS, 2001)

There are additional relevant textbooks and codes including the following.

- *Chemical Process Safety: Fundamentals with Applications, 4th Edition,* (Crowl D. and Louvar 2019)
- *Process Safety Performance Indicators for the Refining and Petrochemical Industries* (for onshore), (API 754, 2016)
- *Process safety – Recommended practice on Key Performance Indicators* (for offshore), (IOGP 456, 2018a)

4.3.2   Hazard Identification and Risk Analysis

Fundamental to process safety is *Hazard Identification and Risk Analysis (HIRA)*. A hazard not recognized cannot be managed nor its risk reduced. The hazard evaluation procedures guideline (CCPS, 2008a) describes many methods and tools, including team selection, data requirements, and documentation as well as the method itself. The more common methods include the following.

- HAZID
- What-if or What-if checklist analysis
- Hazard and operability study – HAZOP
- Failure modes and effects (criticality) analysis – FMEA / FMECA

All these methods are applicable to surface activities (onshore or offshore) and are also applicable and often applied to well hazard identification.

Hazard Identification and Risk Analysis in well construction frequently focuses on barriers. Barriers are a critical component in well construction safety. Refer to Chapter 2 for an introduction into the topic of barriers.

HAZID, What-If, and HAZOP

Refer to CCPS (2008a) which provides full details on these methods. HAZID, What-If and HAZOP are similar techniques using multi-disciplinary teams to identify deviations from normal operations that can lead to a hazardous condition. HAZID and What-If techniques can be applied relatively early in a design before full drawings have been developed. The What-if technique can be aided by the use of a predeveloped checklist. The checklist ensures less obvious but important issues, perhaps identified in past incidents, are considered where brainstorming alone might miss these. The study facilitator ensures that brainstorming is occurring and that the team is not simply working through the checklist. Multiple hazard analyses may be required for the different activities involving numerous specialist groups.

The study minutes are recorded in a tabular fashion, with columns for 1) the what if issue, 2) the identified hazard, 3) the consequence, 4) the safeguards, and 5) any recommendations. As these studies are often carried out early in a design, the findings are used to improve the design and a second more detailed hazard review occurs once a near-final design exists. This design review cycle is reviewed in Chapter 7.

The Hazard and Operability Study (HAZOP) is a powerful hazard identification method that employs a set of guidewords to help the team assess possible deviations. A discussion on HAZOP is included in Chapter 5 for onshore production. HAZOP is not commonly used for well construction operations. Other techniques such as DWOP – Drill Well on Paper are applied. This method is covered later in this Section.

Failure Modes and Effects Analysis (FMEA)

FMEA and a variant FMECA which includes criticality are used for well construction, although primarily for BOP reliability assessments, the results of which directly impact process safety.

FMEA studies start by breaking the system down to main sub-systems. The granularity of the division into subsystems can vary depending on the importance of the subsystem under review. Thus, the whole BOP might be the system and key subsystems might be the annular rams, the pipe rams, the blind shear rams, the kill and choke lines, the control pods, the hydraulic accumulator, the electrical system, etc. Due to its criticality, it might be useful to break the control system down further (e.g. instrumentation, control valves, shuttle valves, solenoids, etc.).

CCPS (2008a) provides generic examples of the failure modes for a normally closed valve. More specific failure modes for a ram BOP might be the following.

- Fail to close ram caused by mechanical problems, seal failure, damaged fittings/tubing, corrosion, improper pipe diameter
- Fail to seal caused by mechanical problems, seal failure, corrosion, material degradation, debris, shavings, improper pipe diameter
- Fail to open caused by mechanical problems, seal failure, debris, hydrates, corrosion
- Fail to shear and seal tubular caused by seal failure, debris, corrosion, incompatible tubular, incorrect space out

For each identified failure mode and its causes, the team assesses the immediate effects at the failure location and the anticipated effects on other equipment and the system as a whole. The typical approach for assessing effects is to identify reasonable worst-case outcomes, often assuming that existing safeguards do not work.

The team documents all safeguards for each failure mode. This is important in assigning maintenance frequencies depending on the criticality of the safeguard and later to support management of change activities.

The team then decides if the safeguards appear insufficient for the hazard and propose any additional safeguards or other actions. This is usually done off-line after the study is completed, as assessing the recommendation can be time consuming and requires approvals.

Minutes are normally in tabular form with columns for the item, its description, failure modes, failure detection, effects, safeguards and any actions required.

When FMEA is applied to BOP reliability studies, it is common to employ the FMECA variant which considers criticality. CCPS (2008a) provides a brief description of FMECA, while ISO 60812 provides fuller details. A prior standard, Mil Std 1629A (1980) now withdrawn, provides a full methodology, and it is readily available and is still used by many. The method is basically the same as FMEA with extra columns added for the criticality assessment.

A weakness of the FMEA method is that it does not consider human factors well. Brainstorming techniques such as What-if and HAZOP are superior in that regard. These methods address both the ergonomics of systems (e.g., display layout, physical effort required) and factors that can degrade human responses (e.g., stress, fatigue, information overload). These can be qualitative or quantitative in approach.

Fault Trees

Fault trees help to understand how complex systems can fail and identify less obvious problems such as common mode failures. The main application of fault trees for well construction relates to BOP operation and reliability prediction. The fault tree method is described in detail in Section 6.3.3.

Risk Ranking Assessment

Risk ranking is an optional additional step in hazard evaluation that can be applied to most methods (e.g., PHA, What-If, HAZOP, FMECA). It is now very common to extend hazard identification to include risk ranking. But for this to be effective, the company should select a single risk matrix for its decision making and not have each project choose its own. This helps with consistency in prioritizing decision making.

Teams examine each scenario and assign a consequence and likelihood level. Many companies have their own risk matrix for this purpose, or they may use the version in ISO 17776 (2016) with six levels of consequences and four levels of likelihood. The number of levels must match the risk matrix being used as the results are plotted onto the matrix. Figure 4-4 shows three decision bands.

- lower risk – manage for continual improvement
- medium risk – incorporate risk reducing measures
- higher risk – fail to meet screening criteria, change required

Other risk matrices employ more or fewer levels (e.g., a 5 x 5 matrix is common) and some matrices include four bands for decision making. Likelihood is often easier for teams to assess when expressed qualitatively as in ISO 17776, rather than quantitatively as in some matrices which specify a frequency band (e.g., 10^{-4} to 10^{-3} per year). The bands provide a consistent basis for decision making for many scenario decisions. Generally, more senior levels of management are involved in making decisions regarding higher levels of risk or where mitigations may be difficult to implement.

The risk matrix shown provides for four types of consequences – to people, assets, the environment, and reputation. Consequences are usually judged on reasonable worst case, but individual companies have their own approaches.

Although the risk ranking approach is simple and easy for teams to understand, there are some disadvantages. Teams may have difficulty selecting the likelihood category if they are not aware of the wider industry historical record. Also, the risk matrix approach applies decision making to one risk at a time. It does not accumulate risk. Thus, many risks all assessed at the lower risk category, when

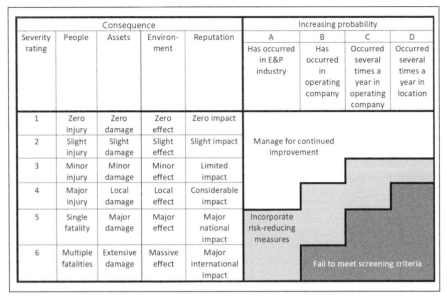

Consequence				Increasing probability			
Severity rating	People	Assets	Environ- ment	Reputation	A Has occurred in E&P industry	B Has occurred in operating company	C Occurred several times a year in operating company	D Occurred several times a year in location
1	Zero injury	Zero damage	Zero effect	Zero impact				
2	Slight injury	Slight damage	Slight effect	Slight impact	Manage for continued improvement			
3	Minor injury	Minor damage	Minor effect	Limited impact				
4	Major injury	Local damage	Local effect	Considerable impact				
5	Single fatality	Major damage	Major effect	Major national impact	Incorporate risk-reducing measures			
6	Multiple fatalities	Extensive damage	Massive effect	Major international impact		Fail to meet screening criteria		

Figure 4-4. Risk matrix
(ISO 17776, 2016) (Permission from ANSI – see Reference for ISO 17776)

considered cumulatively, might add to medium or higher risk. That issue can be addressed by quantitative risk assessment, but this is not common in well construction.

Risk ranking is very common in the upstream oil and gas business and is used across all levels and types of activities from corporate risk to project risk including drilling projects.

Drill Well on Paper – DWOP

The Drill Well on Paper approach (DWOP) is a method developed by drilling contractors and a short description is available from Halliburton (2015). The methodology for both onshore and offshore wells uses a mixed group of senior personnel and technical specialists, both office-based and facility-based, from the operating company and its contractors. The method is similar in concept to a What-If study in that it is free flowing brainstorming.

At the time of the DWOP study, most design and technical alternatives have been evaluated and a final well design selected. The DWOP is a final check before drilling commences. In addition, DWOP can address efficiency issues as well as risks.

DWOP analyzes each step of the well construction process seeking to improve performance and reduce risks and costs and assure the safe construction of the well (e.g., by elimination of non-productive time). The team examines in detail the well drilling plan step-by-step to ensure it meets regulatory or code requirements or best practices. The team results are combined and discussed until consensus is reached.

Past optimum well drilling performance becomes a target for the new well and helps to include lessons learned.

Bow Tie Analysis

The bow tie analysis is a newer method which has a focus on explaining and demonstrating risk management barriers rather than being a tool for decision making directly. It is part of the RBPS element *Hazard Identification and Risk Analysis*. The bow tie method is described in CCPS (2018c) and briefly in this book in Section 2.7. An example bow tie diagram is shown in Figure 2-11.

Bowties are helpful in providing a visual representation of the barriers that prevent or mitigate a risk. They provide an overall view for those charged with managing risks. They also provide a specific view of how the various barriers eliminate or mitigate risk which allows the individual operator or maintenance technician to see how their work supports risk management.

The bow tie is not a decision tool directly as it does not quantify the risks of each arm, but it does show where there might be a lack of barriers (e.g., a pathway with only one barrier) or an excess number. A key use and benefit of the bow tie is to identify where and what type of barrier health data should be collected and on what frequency.

Bow tie diagrams are created in team sessions that start with the outcome of some prior hazard evaluation (PHA, What-If or HAZOP) combined with risk ranking. These generate the higher risk scenarios and the more important scenarios can be selected for bow tie creation. The hazard evaluation minutes contain a column titled safeguards which contains a mixture of barriers (full IPLs) and degradation controls supporting those barriers. The facilitator collects the team results and constructs the bow tie diagram, usually using commercial software (CCPS, 2018c provides several examples).

An example bow tie diagram was shown earlier in Figure 2-11. Bow tie diagrams get complex if there are many different threat arms and software can display or hide barrier decay mechanism arms as appropriate to make diagrams easier to understand (Figure 4-5) (CCPS, 2018c).

4.3.3 Asset Integrity and Reliability

Developing a Well Integrity Plan is key for well construction. The well integrity plan is similar to the *Operating Procedures* and *Safe Work Practices* elements in RBPS. Well integrity is presented in API RP 100-1 for onshore fracking wells, Norsk O&G (2012 and 2016) for offshore wells, and ISO 16530-1:2017 (well integrity for the lifecycle) for all types of wells, and by company internal well integrity guidelines. Well integrity both onshore and offshore is addressed in separate API standards which collectively deliver the required well integrity.

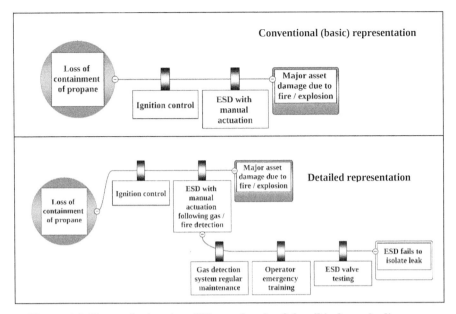

Figure 4-5. Example showing different levels of detail in bow tie diagrams

Two useful definitions of well integrity are as follows.

API 100-1 (2015b): "The quality or condition of a well in being structurally sound and with competent pressure seals (barriers) by application of technical, operational, and organizational solutions that reduce the risk of unintended subsurface movement or uncontrolled release of formation fluid."

NORSOK D-010 (2013): "Application of technical, operational and organizational solutions to reduce risk of uncontrolled release of formation fluids throughout the life cycle of a well". Also, "There shall be two well barriers available during all well activities and operations, including suspended or abandoned wells, where a pressure differential exists that may cause uncontrolled outflow from the borehole/well to the external environment."

The first part of D-010 is the same as API 100-1. The D-010 requirement for two barriers at all times is not present in API; however, the BSEE Final Drilling Rule for offshore drilling does require this when removing well control equipment. Note API 100-1 applies to onshore wells, API RP 65-2 (API, 2010) and API 96 (API, 2013a) address offshore well integrity (and some use for onshore wells), along with several other API standards.

BSEE (2016) specifies how to achieve well integrity in its Final Drilling Rule, including all necessary permits. This rule does not apply to onshore wells which follow API standards, such as API 100-1.

The core industry documents on well integrity include ISO 16530, NORSOK D-010, and UK Oil & Gas Well Integrity guidance. They identify critical elements for well integrity that should be included in a well integrity plan. Norsk O&G (2012 and 2016) identify the detailed knowledge of well construction and barrier functionality for well integrity and how this information should be organized into well handover documentation (i.e., drilling contractor to operating company). Well barrier schematics are helpful to identify the primary and secondary barriers and all the barrier elements. These schematics also include details of casing, cement and mechanical barriers, and information on formation strength from data collected during drilling.

A well integrity management system maintains well integrity through all phases of well life. Key topics include technical standards employed, barrier definitions, equipment requirements, and safety systems. Operational procedures define maximum loads of pressure, temperature, flowrate, and corrosion causing aspects. Monitoring programs confirm these conditions and emergency plans define response to a loss of well control or degraded barriers causing sustained casing pressure. The well integrity management system can be integrated into SEMS or RBPS, but this would be spread between several elements, so some companies choose to maintain a separate well integrity management system that brings everything together in one place.

4.3.4   Training and Performance Assurance

The RBPS element *Training and Performance Assurance* is especially important for well construction as human performance is key to recognizing a potential kick event and selecting an appropriate response. Kick events can be hard to identify and can be masked by other activities. IADC (2015c) provides some indicators that a kick may be occurring. The list below is a selection.

- Mud flow out of the well greater than flow rate into the well
- Positive pressure on the drill pipe when not pumping
- A change in drilling penetration rate (higher or lower)
- A decrease in pump pressure
- A sudden torque increase
- A change in mud properties
- A change in cuttings characteristics

None of these are a direct indication that formation fluids are flowing into the well; instead they indicate that conditions are changing that may signal an increased risk of a kick. Some offshore drilling operations have, in addition to the driller, mud loggers and real time monitoring centers onshore. Both use a large array of data measurements and computer graphical displays to assist the driller in determining if there is a well influx.

As well as indicators, IADC also outlines multiple possible responses. These include shutting in the well using the annular or ram preventers or circulating out the kick with initially current and subsequently a heavier mud weight (the "drillers method").

Given that kick detection and response is not automated and requires human intervention, training and competence assurance is vital. Regulations for offshore require certified training of personnel on well control and testing. This training and testing includes the use of well control computer simulators where many multiple scenarios of kick detection and well kill operations are practiced. However, training alone does not address all the issues as even well-trained humans make mistakes. Companies must understand factors that lead to human error and try to eliminate these. The topic of human factors is beyond the scope of this book and is addressed in CCPS texts (2004 and 2007b) as well as many other textbooks.

4.3.5 Emergency Management

If kick detection and response fail, then there is the potential for a loss of well control and necessary emergency response. This links to RBPS elements on *Operating Procedures* and *Emergency Management*. IADC (2015c) reviews the full range of well control and response measures. Important responses to a well kick include detection and "circulating out" the kick using heavy drill mud and closing the BOP if this has not already been done. If the incident continues and a loss of well control occurs, the driller can actuate a diverter valve to direct any well fluids away protecting the locations where crew are located. Alternatively, offshore the MODU can disconnect and sail away to a safe location.

Once the rig and its personnel are protected, then environmental responses are necessary to contain and collect any well fluids. Onshore these might include building temporary containment barriers and pumping out to collection trucks. Offshore measures include booming, dispersant injection and skimming, as well as well capping and subsea containment operations with special purpose capping stacks. If well capping is not successful, then it will usually be necessary to drill a relief well to intercept the flowing well and inject cement to terminate the flow. An outline of a full range of environmental response measures is included in Shell (2015) in its public submission to BSEE for drilling off Alaska.

The Deepwater Horizon blowout event showed that the offshore industry needed more resources for the largest scale of loss of well control incidents, more than any single company could provide. Three consortia were created, two in the US and one globally, to acquire and maintain equipment and people ready for immediate deployment. This equipment includes a semi-submersible drill rig and a floating production facility, booms, response boats, capping stacks, etc. These facilities, after installation, are capable of trapping and storing oil before it reaches vulnerable ecosystems.

While well construction activities have personnel formally trained in well control and blowout prevention, well blowout events often require specialist well control companies to assess and terminate the event.

4.3.6 Learn from Experience

Learn from Experience is one of the 4 pillars of RBPS. This pillar includes *Incident Investigation, Measurement and Metrics, Auditing,* and *Management Review and Continual Improvement.* These tools are also in API RP 75 SEMS which is required by regulation in the US OCS. CCPS (2008b and 2019c) has published texts on incidents that define process safety. While this is very useful for periodic safety talks and to build process safety knowledge in newer personnel, this primarily focuses on downstream incidents, although it does include the Deepwater Horizon incident.

Effective hazard identification requires that teams identify all potential process safety events. Most companies have internal communication systems to share incident lessons. However, process safety incidents are rare and may not have occurred within the facility's experience.

There are at least four important mechanisms to help companies ensure that personnel are aware of these rare yet important incidents.

1. Regular reviews of industry incidents – In the US, state regulators report on onshore drilling incidents, and the CSB (2019) has started to investigate onshore drilling incidents. BSEE and other offshore international regulators also report major incidents both on their own websites as well as a consolidated list on the International Regulators Forum website. API RP 754, and its offshore equivalent IOGP 456, provide definitions of leading and lagging indicators for four tiers of process safety events. Companies are starting to follow these guidelines and are reporting publicly on the more serious Tier 1 and 2 events. COS provides a listing of incidents. Further details were provided in Chapter 1. It is the role of process safety specialists and design engineers in the company to monitor such statistics and provide safety talks or updated company design rules to address such incidents.

2. Participate in and employ current codes and standards – Engineering bodies (e.g., API, ANSI, IADC) update their codes and standards periodically and these updates address any important incidents if the existing documents do not address the issues adequately. Companies that participate in these committees get early notice of changes and, by interacting with other company specialists in the committee, learn of incidents that may not be published.

3. Participate in industry conferences and public meetings – CCPS, SPE, API, COS, IADC and other bodies organize periodic conferences which address technical advances, upcoming standard updates, and often provide recent incident summaries. Participating in these events helps process safety and well construction specialists keep up to date on technology and aware of industry incidents.

4. Training in process safety – Process safety training is available from industry associations including CCPS. The CCPS offers Safety and Chemical Engineering Education (SAChE) courses focusing on university students addressing process safety and the RBPS system (SAChE, 2019).

Some universities recently have begun to include process safety in their curricula. The upstream industry has addressed process safety in detail in its own training courses since Piper Alpha.

4.3.7 Management System Audits and Safety Culture Surveys

Onshore and offshore facilities apply safety and environmental management systems to control their own and their contractor activities. Audits are an essential aspect of these management systems to ensure that what is specified actually occurs. These management systems apply to all aspects of upstream operations, not only well construction. In the US, larger onshore facilities usually follow PSM (OSHA 1910.119), while offshore, BSEE mandates SEMS which is based on API RP 75 with several additional requirements. Both PSM and SEMS require audits. BSEE requires periodic third-party independent audits. The Center for Offshore Safety has developed SEMS audit requirements (COS, 2014) and an audit service provider accreditation system that help to ensure effective and consistent audits.

Auditing and *Management Review* are elements in the RBPS pillar *Learn from Experience*. Audit results are considered in the Management Review, which also considers safety and environmental performance, incident investigation results and learnings, and mechanical integrity statistics relating to important barriers, to decide whether any changes to the current management system are warranted. The process of measuring current results (whether by metrics or by audit) and addressing these with changes if performance falls below target demonstrates continual improvement.

Safety culture is a recognized issue in both the downstream (Baker, 2007) and upstream industries (Deepwater Horizon Commission, 2011). *Process Safety Culture* is the first element of RBPS. The application of PSM or SEMS helps create a positive safety culture. However, this is not a guarantee that a company will reach the high level of process safety culture that it desires. The Baker report differentiates between a general safety culture (which focuses on more frequent occupational safety risks) and a process safety culture (which addresses rarer major incident risks). It is possible to be excellent in the former but weaker in the latter, as was seen in the Deepwater Horizon incident. The introduction of SEMS in the US and safety case elsewhere have the intent to improve safety culture offshore.

The usual tool for assessing culture is a survey. This can be a written questionnaire or a series of focus group meetings, both have advantages and disadvantages. The questionnaire has the advantage of anonymity and may be completed by everyone. But since it is generic, its responses tend to be general (e.g., "my supervisor values production more than safety"). Focus groups are not anonymous and may be impractical to apply to all personnel, but they do typically identify specific instances that are addressed more easily (e.g., "at the last shutdown, contractors started working without fully following permit requirements"). The Baker Panel (2007) provides an example survey questionnaire and shows how this is scored.

BSEE (2013) has published its policy covering nine characteristics contributing towards a positive safety culture, based on similar guidance for the US nuclear industry. The nine characteristics are as follows.

1. Leadership commitment to safety values and actions
2. Hazard identification and risk management
3. Personal accountability for process and occupational safety
4. Work processes address safety and environment
5. Continuous improvement for safety and environment
6. Positive environment for raising concerns
7. Effective safety and environmental communication
8. Respectful work environment
9. Inquiring attitude avoiding complacency

COS (2018) has issued guidance on offshore safety culture and has a SEMS maturity self-assessment tool. The COS safety culture good practice covers six of the characteristics above, as the other three topics (hazard identification and risk management, work processes, and continual improvement) are covered sufficiently through API RP 75.

5

Application of Process Safety to Onshore Production

5.1 BACKGROUND

Once a well is constructed and production is determined feasible, then a production system is developed as shown in the life cycle in Figure 2-1. Alternatively, if production exists nearby, the new well may be tied into that production system. The process safety activities during the design activity are covered in Chapter 7.

Onshore production scale varies from single low-pressure producing wells (e.g., pumping units), through several wells flowing to manifolds and then to medium size integrated production facilities to very large integrated facilities with enhanced oil recovery (e.g. Alaska, Middle East, etc.). There are process safety risks associated with all these facilities.

Onshore facilities have process safety advantages and disadvantages compared to offshore facilities. Advantages include greater spacing potential between hazardous items and occupied buildings, remote personnel accommodation (e.g., either commute daily to the site or live in well separated accommodation modules), maintenance and inspection activities have better access, and less congested laydown areas. The main disadvantages are that onshore facilities potentially impact public areas, neighboring agricultural activities (ploughing or digging) can impact gathering or export lines, and spills can end up in local waterways. While not a process safety issue, onshore facilities can cause noise, odors, and traffic hazards.

In many respects, onshore production process safety issues are not so different to downstream processes of similar scale. There are leak sources from production equipment and storage vessels, which can lead to fires, explosions, or toxic impacts to personnel and any communities within the hazard zone. An examination of Marsh's *The 100 Largest Losses in the Hydrocarbon Industry* (2020) shows that historically there are few large-scale onshore events, such as the Longford incident described below, as compared with offshore. However, there are a number of lesser scale safety and environmental incidents which are assessed by various state regulators in the US (e.g., Colorado Oil & Gas Conservation Commission and similar in other states). In the UK, the only significant onshore oil and gas development is at Wytch Farm and the HSE maintains incident statistics. This is the largest onshore field in Europe. Incident data in Australia is available from the individual state regulators.

Incident: Longford Gas Plant, Australia 1998

Longford is an onshore gas plant servicing the Bass Strait fields off SE Australia. Personnel were trying to rectify a heat exchanger that had become covered in ice. The exchanger ruptured and a subsequent fire killed two and injured eight more personnel. A long-lasting fire occurred that ultimately resulted in the loss of gas supply to the State of Victoria for over two weeks with major disruption to the public and industry.

Process Safety Issues: A change in well fluid composition occurred over a long period of time and resulted in a separation column eventually allowing propane to reach parts of the plant not designed to receive it. This part of the plant operated at low pressure and the propane flashed to -40°C and this caused a heat exchanger to ice over. Operators and maintenance personnel had never experienced this type of event and were unaware that such temperatures cause embrittlement to normal steel. They introduced hot oil into the exchanger and the rapid temperature rise caused internal stresses and the exchanger end failed catastrophically. The site wished to avoid a full loss of gas supply to the State so a total shutdown was to be avoided if possible. With the fire at a major pipe intersection and no good drawings to use, responding personnel resorted to walking down many lines to try to isolate the leak. Ultimately this failed and the plant was shut down.

There was a lack of knowledge by the local personnel of embrittlement issues for mild steel and the personnel who knew had been moved to Melbourne and were not immediately available for advice. Also, drawings needed for emergency isolation following the incident had not been kept up to date and this hindered the response.

Source: Longford Royal Commission (1999)

RBPS Application

Process Knowledge Management: Personnel operating equipment that can potentially reach temperatures that cause embrittlement need to be trained in those hazards and how to respond to them. The operator argued that knowledge of this type of hazard was only becoming known at the time of the incident.

Management of Change: Personnel who might have been knowledgeable in the embrittlement issue were not immediately available for advice. This incident was a key driver in the development of the *Guidelines for Managing Process Safety Risks During Organizational Change* (CCPS, 2013b).

Another aspect of *Management of Change* is that isolation of the leak source, while maintaining plant production, required that drawings be kept up to date, and this had not been done.

Typical Production Facilities

Two example facilities are shown in Figure 5-1 displaying the range from simple to the more complex. The first shows a smaller-scale onshore oil production facility and the second a large-scale onshore production facility in Alaska. The medium-scale facility has separation on the well pad with holding tanks for liquids and export by pipeline. The large-scale integrated facility is housed in a building to provide protection against cold weather and allow for routine operations, inspection and maintenance in a protected environment. It exports liquid product via pipeline and produced gas is reinjected back into the reservoir as there are no gas export facilities.

The range of production activities depends on the source hydrocarbons and the intended products.

Figure 5-1. Example onshore smaller-scale production and large-scale production facilities

Gas Processing

Gas processing can be considered either upstream or midstream and is covered briefly here. Gas to be exported into the commercial gas pipeline network must be processed to achieve inlet typical criteria (Wichart, 2005) for the following.

- Hydrocarbon dewpoint limit
- H_2S and total organic sulfur limits
- Water content limits
- CO_2 limits
- Minimum heating value (the Wobbe Index)

A simplified diagram for overall well fluids treatment was shown in Chapter 2 (Figure 2-10). Treatment would start with separation, often in two stages, giving gas, liquid, and water phase streams. Gas is compressed and sulfur (if present) is removed in a sweetening plant often by lean amine absorption, although many other solvents are also used. The rich amine is heated and evolves the absorbed sulfur products, and these are sent to a sulfur plant that makes liquid sulfur. Dewpoint control aims to remove hydrocarbon condensables (C_3+) that might condense in a long export pipeline and create liquid slugs. It also dehydrates the gas for similar reasons. Dehydration is achieved often using glycols which are regenerated when saturated. The product from dewpoint control is a commercial sales gas.

Separated liquids require further treatment. Hydrocarbons are recovered by distillation into LPGs (C_3 and C_4) and into a heavier stream of C_5+ materials. Separated water is collected and requires treatment as it is saturated with hydrocarbons. This may be done locally or sent off for treatment. Some facilities also reinject the water into the reservoir.

Produced gas may be converted into LNG in large scale liquefaction plants. This typically happens at coastal port locations as the LNG is exported in large ships as it is only practical to pipeline cryogenic LNG over short distances. Since these are not part of onshore processing facilities, they are not covered here. However, offshore FLNG facilities are linked to the wellhead and are covered in Chapter 6.

Oil Processing

In wells that are primarily producing oil, the processing is usually simpler than in gas plants. This is because the oil is not for direct use but is exported to a refinery where further processing to commercial sales products occurs. The main processes are water separation and degassing so that it may be shipped safely. The Lac Megantic rail incident in Canada (Lacoursiere et al, 2015) was partly due to inadequate degassing at the source facility that allowed a nominally heavy crude to become pressurized during transit. Many tank cars exploded when the train derailed (see later example incident box for further details).

Produced oil is either sent immediately to pipeline for export or is stored onsite in atmospheric storage tanks for export by road, rail or ship. There can also be temporary storage for off-specification material for reprocessing.

5.2   ONSHORE PRODUCTION FACILITIES: RISKS AND KEY PROCESS SAFETY MEASURES

The identification processes for risks are generally one of the first activities carried out. The actual processes used are summarized later in Section 5.3.2.

5.2.1   Leak from Production Facilities

Risks

There are multiple potential leak sources at upstream gas plant facilities, where most of the production is carried out at moderately high pressures that typically exceed 1000 psi (69 bar). This means that even small holes can cause large leak rates. Typical leak sources include pipes, flanges, small bore connections, compressor and pump seals, and vessels. Corrosion is often a contributory cause as well fluids contain water and acid gases. Other causes include vibration from compression activity, erosion from sand in the oil, impacts from work activities, or design or construction defects (e.g., incorrect gasket materials).

A leak of flammable materials can lead to jet fires and/or pool fires. If a vapor cloud forms and is subsequently ignited, a flash fire occurs. If sufficient congestion exists, the flash fire flame can accelerate and cause a vapor cloud explosion. Means to estimate potential outcome hazard zones are provided in the *Guidelines for Chemical Process Quantitative Risk Analysis* (CCPS, 1999). The actual outcome for a leak depends on the initial condition and how the event progresses. This is shown in Figure 5-2 from IChemE (1996). This is complex and not easy to execute using manual techniques and most companies use specialized consequence software.

As mentioned previously, some onshore production facilities are located inside large buildings, for example on the North Slope of Alaska or North Africa, which allows operators and maintenance personnel to work in a temperate environment which enhances process safety. A downside to locating large facilities inside buildings is that leak events, which safely disperse with the wind if outside, can more easily create flammable clouds inside. This increases the potential for flash fires or vapor cloud explosions in congested spaces (see box Vapor Cloud Explosion – Short Primer in Section 5.3.2). This hazard is managed by preventing leaks (e.g., mechanical integrity program) or mitigating them if they occur through the provision of gas leak detection with ESD and a more extensive fire detection and suppression system than is typically used in outside locations.

Sour gas fields contain H_2S and organic sulfur compounds. In the US, H_2S concentrations are usually less than 4%, but are much higher in wells in several states (e.g., East Texas, Colorado, and Wyoming). In the Middle East, some fields are 50% H_2S. A leak of produced gas at 4% (40,000 ppm) H_2S is a serious toxic hazard as the ERPG-3 is 100 ppm (the concentration that most people can survive a 1 hour exposure) and concentrations above 500-1000 ppm can be rapidly fatal. This is mitigated to some degree if the vapor produced is buoyant (i.e., dominated by methane content) that tends to disperse any leak upwards, away from ground level. However, natural gas with significant C_3+ and H_2S can be a dense gas and not

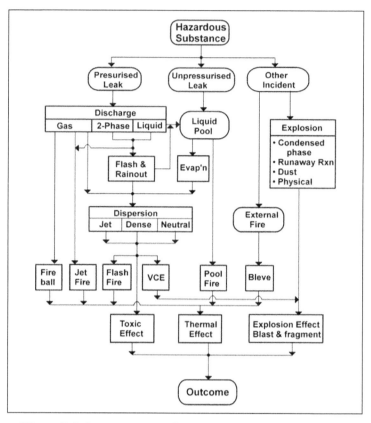

**Figure 5-2. Source term pathways to ultimate consequences
(IChemE, 1996)**

buoyant. A greater hazard is associated with the treatment process where the regeneration of rich absorbent liquid produces a nearly pure H_2S stream. This is immediately sent to a sulfur plant that converts H_2S to pure sulfur or injects the acid gas back into the ground. Some facilities use double pipe for added safety for this connection.

The toxic criteria used by the EPA for assessing hazardous facilities is ERPG-2 (Emergency Response Planning Guideline – Level 2), also known as AEGL (Acute Exposure Guideline Levels). This is the concentration that most people can be exposed to for one hour without developing life threatening symptoms. The ERPG-2 for H_2S is 30 ppm. H_2S leaks initially have an associated rotten egg odor, but after a short time olfactory fatigue causes a loss of the sense of smell and people can be exposed to toxic concentrations without recognizing it. This is a particular hazard to personnel who are close to the source and the neighboring community since toxic clouds can travel significant distances.

Great care is attached to protecting the environment and wildlife in Alaska as is seen in this common scene of a caribou grazing beside a pipeline in a North Slope production facility (Figure 5-3).

Key Process Safety Measure(s)

For leaks, all the elements of RBPS, SEMS and PSM are important, but two are especially relevant and are listed below.

Hazard Identification and Risk Analysis: Through the use of HIRA tools, a better understanding is obtained regarding the precise nature of the hazards presented by leaks from production equipment. This helps facility designers identify the proper safeguards to prevent or mitigate leak events. These include good operating procedures and safe work practices, inspection and maintenance, flammable and toxic gas detection, emergency shutdown systems, consequence calculations to estimate potential hazard distances, and area classification to reduce ignition likelihood.

Emergency Management: Control rooms on larger facilities not only protect personnel but serve important roles in emergency response. Modern practice is to either separate the occupied building from the hazards or to make the control room resistant to the types of consequences identified in the HIRA. Guidance is provided in API 752 and 753, with additional details in CCPS (2018b). Fire protection measures are summarized in CCPS (2003) with additional guidance in NFPA and API Standards.

Figure 5-3. Caribou grazing by pipeline in production facility

5.2.2   Gathering Pipeline Leaks

Risks

Many larger onshore production facilities are located some distance away from the wellheads. Gathering pipelines collect the raw well fluids for processing in the facility. Fluids may be three phase – hydrocarbon gas and liquid, and water. Return lines may send gas or water back to the well site for lift operations to increase well production.

Leaks are possible from these lines. The raw fluid may be corrosive containing acid gases as well as produced water. A credible scenario of note is the impact of these lines by third parties via unverified digging. Buried lines in remote areas may be hard to inspect, but tools such as leak detection systems and aerial inspections / drones can assist in finding leaks once loss of containment occurs. In cold climates lines are not buried due to problems with permafrost but instead are mounted on supports. For these elevated lines, wind-induced vibration can lead to pipe metal fatigue and leaks. A creative solution is the use of weights attached to each span of pipe to damp out these vibrations (Figure 5-4).

Leaks cause safety issues as well as environmental impacts. In remote locations, leaks may not always be identified immediately, and the potential exists for environmental impacts if the leak gets into local watercourses or other sensitive environmental receptors. Depending on routing of the pipelines, local communities may also be affected.

Figure 5-4. Weights attached to pipe to dampen vibrations

Key Process Safety Measure(s)

Asset Integrity and Reliability: Designing equipment for both the process and the environmental conditions is important during design stage. Thereafter, inspecting and maintaining integrity protects against leaks over the life of the equipment.

Stakeholder Outreach: Where a leak may impact external parties, it is important to reach out to them in advance of any incident. This helps to reduce the likelihood of inadvertent pipeline damage and helps mitigate any incident damage through advanced planning. Appropriate signage and physical guards are important to prevent accidental damage to pipelines.

5.2.3 Storage Tanks and Vessels

Risks

Produced condensate and oil, once stabilized (i.e., degassed), are transferred to atmospheric storage tanks until exported. Onshore, tanks are typically sited inside dikes sized to contain 110% of the largest tank volume. Generally, atmospheric storage is low risk with the primary hazard being a rim-seal fire (more likely), but potentially a full surface fire can occur (much less likely). Leaks from connections or the export pump accumulate in the tank dike. However, in certain circumstances of tank design, ambient weather, hydrocarbon characteristics and a congested space, a tank overflow cascading to ground level can generate a vapor cloud and, if ignited, can explode as occurred in Buncefield UK in 2005 and Jaipur India in 2009. These were both gasoline tank events, but stabilized oil sometimes has a similar vapor pressure, and thus a potential for similar outcomes.

Atmospheric storage tanks are more complex designs than a simple cylinder with a floating roof. The floating roof of an atmospheric storage tank is vulnerable and subject to sinking if there is corrosion of the buoyancy chambers or if accumulated rainwater is not periodically drained away. A jointed connection from the center of the roof is used to take collected rainwater and discharge it at the side of the tank near the bottom and into the storm water system. If the valve is left closed, then too much water can accumulate on the roof and sink it. Alternatively, if the valve is left open, a leak at the internal joint can drain oil into the dike. Either event creates a serious hazard. The primary protection against these events is a trained and competent workforce to judge when valves are to be opened and closed.

Cone roof tanks are also employed on small facility well pads and where the product stored is volatile. These rely on the vapor space being above the upper flammability limit (UFL) to make the vapor space safe from ignition. Inert gas systems are also used to make the vapor space safe.

Atmospheric tanks have several weak points that can lead to loss of containment. The shell bottom seam is frequently subject to corrosion causing a leak. Also, inlet and outlet connections are known leak points. A regular inspection program to monitor pipe and shell wall thickness and operator patrols to quickly spot leaks are the most common control measures. Vacuum failures, due to blocked vents, for example, are also common and can lead to loss of containment incidents.

Incident: Lac Megantic Fire and Explosion event, 2013

A 74-car train transporting oil from the Bakken formation to the Saint John refinery in Canada was parked on a hill above the town of Lac Megantic, Quebec, and left unattended. Loss of compressor power for powered brakes and inadequate setting of train manual brakes allowed the train to roll into the town below where multiple tank cars derailed and exploded. A subsequent fire caused several more rail cars to BLEVE causing 47 fatalities and destroying 30 buildings.

Process Safety Issues: While this incident occurred in the transportation phase of the oil's journey, a contributing cause was that the onshore production facility had not stabilized the oil sufficiently. The dissolved gases came out of solution during the trip and the high local ambient temperature pressurized the rail cars. In the derailment, several rail cars were punctured and caught fire, and several resulted in BLEVEs, causing the catastrophic outcomes.

Many of the mitigations were transport related (e.g., to strengthen rail cars and avoidance of unattended trains above populated locations).

Source: Transportation Safety Board of Canada (2014)

RBPS Application

Process Knowledge Management: A key aspect of this incident was understanding the material properties and managing them appropriately; specifically, adequately stabilizing the oil before transport.

Operating Procedures: While we tend to think of procedures as applying to process operators, in this case the train driver did not follow correct procedures. An insufficient number of manual brakes were set when the train was left unattended, and a fire in the boiler chimney later caused the engine to shut down and stop the compressor, releasing the normal braking system. The manual brakes were unable to stop the train running down the hill into the town.

Line purging can cause a process safety hazards to tanks as occurred at a Sonat well fluids processing facility in 1998 (CSB, 2000). That incident occurred due to an error related to 11 valves that needed to be manually repositioned. Two key valves were incorrectly closed and purge gas consequently over pressured a tank, which burst and caused 4 fatalities. The CSB identified issues relating a lack of a formal design review, a lack of written procedures for the task, and inadequate employee training.

If a gas plant recovers LPG as a product, then this may be stored in a pressure vessel until it is exported. LPG is a pressurized liquid at ambient temperature and leaks propagate as shown in Figure 5-2. This can give rise to jet and flash fires, vapor clouds, and the potential for vapor cloud explosions. A special hazard of interest involving LPG tanks is that of BLEVE, Boiling Liquid Expanding Vapor Explosion. This is usually caused by an external fire that can cause the upper portion of the tank shell to weaken and fail catastrophically. The NFPA has been active in alerting firefighters to this potential and of appropriate response measures. In upstream facilities it is common to install water sprinklers on pressure vessels that hold large inventories. The aim is to keep the vessel shell cool until a fire is extinguished. This protects against BLEVE events.

Key Process Safety Measure(s)

Process Knowledge Management: As in the Lac Megantic incident, it is important to understand the properties of the materials being handled and use the correct metallurgy for the equipment, piping and other accessories.

Conduct of Operations: While both the Buncefield UK and Jaipur India incidents were noteworthy, they both started with a tank overfill. Paying attention to operations and conducting them appropriately each and every time helps to prevent significant incidents. Also, as in Buncefield, when a critical instrument (here a level gauge) signals a problem, this should be addressed.

5.2.4 Loss of Well Control

Risks

There are several activities during the production phase that can result in a loss of well control. These are primarily interventions and workovers to maintain or enhance reservoir production. Refer to Chapter 4 for discussions on these topics.

A potential cause for loss of well control, onshore or offshore, is a wellhead mechanical impact or corrosion failure that allows well fluids to escape. The barriers protecting against this event are bollards, the Christmas tree installed on the wellhead, and a subsurface safety valve (SSSV) where one is installed.

Another production activity that can result in a loss of well control is associated with gas or water lift activities. These production enhancement procedures involve sending gas or water into the annular space and then injecting this into the production tubing near the bottom of the well. A loss of hydraulic head or gas pressure, for example due to surface line damage near the well head, can result in reverse flow up the annular space and a loss of containment event. The Christmas tree valve is designed and installed to close off this reverse flow potential. Subsurface safety valves, if installed, apply to the production tubing and not the annular space, and they are not a barrier for this event. For the subsurface safety valve to be effective in a gas lift operation it must be set below all the gas lift injection points in the tubing. This is seldom practical.

Key Process Safety Measure(s)

Asset Integrity and Reliability: Arranging and protecting wellhead equipment to minimize the likelihood of mechanical impact aids in preventing loss of well control. Inspecting for and addressing any corrosion issues through a corrosion control management system reduces the likelihood of these events resulting from corrosion failures.

5.2.5 SIMOPS

The hazard of simultaneous operations was previously mentioned in Chapter 4 and is covered in detail here.

Simultaneous operations refers to two or more distinct activities occurring at the same time and in close proximity to one another. Well construction and production are two distinct activities and are thus considered SIMOPS. Project work during production is also SIMOPS. Conversely, producing from two adjacent wells are not distinct activities and are not considered SIMOPS. The term combined operation is also used, but IADC notes they are different, with combined operation normally meaning two separate facilities working adjacently (e.g., a drilling rig beside a production facility). Regardless of the term used, this is a SIMOPS activity and requires appropriate controls.

While each individual activity may be managed safely, the potential for interaction between the activities significantly increases the risk unless the interaction is actively managed. Guidance is available for SIMOPS during well construction from API Bulletin 97 (API, 2013b), which provides guidance on the interface between an operator's management system and the drilling contractor's safe work practices (see also Section 4.3.1). While these have an offshore tone, the principles of an adequate operator-contractor interface are equally applicable onshore.

While API 97 is designed for well construction and production activities, the framework also works for project activities adjacent to production or well construction. The hazards are different, as are the necessary control measures, but the active management of the interface is similar.

Key Process Safety Measure(s)

The key process safety measures are typically the elements of an interface document including the following.

- A statement of management system principles
- An overview of the two or more activities
- *Hazard Identification and Risk Analysis* including key safety barriers
- *Safe Work Practices* including permits
- *Management of Change*
- *Conduct of Operations* including a responsibilities chart (e.g., RASCI)
- *Emergency Management*

5.2.6 Vents and Flare

Risks

There is a risk of thermal and toxic exposure to material from vents and flares. Onshore production facilities normally have a flare to burn off any emergency release of flammable or toxic vapors. The flare collects releases from pressure relief valves on process vessels, except on small well pads which may employ atmospheric reliefs. When a large release is flared, this may create a thermal hazard at ground level. Common practice is to surround the flare with an exclusion zone based on a specified maximum radiation exposure (e.g., API 521).

Flaring a gas containing H_2S does not eliminate its toxic potential as the product of combustion, SO_2, is as toxic as H_2S. However, this is contained in hot combustion gases that rise and its concentration dilutes as it disperses downwind. Ground level concentration calculations should be carried out to determine if the flare height is sufficient to ensure safe concentrations of SO_2 at any location occupied by the public. Depending on the SO_2 concentration, the flare height may be higher than that required for thermal exposure protection.

Another hazard associated with flaring includes a flame-out event which could allow flammable or toxic vapors to reach ground level or elevated work locations. Less likely, but possible, are hot embers from the flare tip causing ignition downwind. API 521 requires that this event be assessed and is a factor in determining flare height and location. The flare knockout pot collects waste streams with volatile liquids that can be corrosive and should be designed and sized appropriately to prevent liquid carry over and release from the flare tip.

Key Process Safety Measure(s)

Process Knowledge Management: Vents and flares should be engineered based on the material properties and volume with a design intent to minimize risks to nearby personnel and to protect process equipment connecting to the flare.

Hazard Identification and Risk Analysis: Consequence calculations are used to determine ground level thermal radiation or concentrations of toxic materials and these can be used to set exclusion zones, or to assist with *Emergency Management* planning.

5.3 APPLYING PROCESS SAFETY METHODS IN ONSHORE PRODUCTION

There are a wide range of process safety methods and applications that are employed for onshore production facilities. This section reviews some of the more common ones.

5.3.1 Regulations and Standards

Standards

There are a large number of API and ISO/IEC standards relevant to onshore production. API (2015a) lists over 100 safety standards for onshore facilities, including onshore drilling. API standards are complemented by other standards, such as ASME VIII for pressure vessels (American Society of Mechanical Engineers), a range of NFPA standards for fire protection (National Fire Protection Association, e.g., NFPA 30), and guidance on corrosion protection from NACE (National Association of Corrosion Engineers).

Smaller onshore oil facilities tend to follow API and related standards (listed above), and either RBPS or the OSHA PSM regulation (although they may be exempt formally) for structuring their process safety management system.

Regulations

An overview of regulations was provided in Section 2.8. Specific to onshore production, OSHA and EPA share responsibility in the US for onshore regulations for facilities exceeding nominated inventory thresholds. Similarly, PHMSA regulates pipelines. OSHA PSM has a focus on worker safety, while EPA RMP (Risk Management Program) has a focus on offsite / public safety and environmental incidents.

OSHA does not require a single document collecting all elements of the PSM program, just that all elements must be present. Conversely, the EPA does require a risk management program document.

For facilities in Europe and parts of Australia (e.g., Victoria), a safety case must be prepared. The EU regulation is known as the Seveso Directive, originally passed in 1986, and updated twice to address serious incidents that were not well covered. This regulation was covered in Section 2.8.

For larger onshore production facilities that exceed regulated thresholds, both the EPA RMP and the safety case (now known as safety report) requirements in the EU Seveso III Directive require that important parts of the major hazard plan be documented. This includes a listing of hazardous materials, the facility description, hazard identification and risk ranking, the management system, and the emergency response plan. In the US, most states have regulations for production facilities.

Both RMP and Safety Reports must address emergency response and this topic is outlined in Section 5.3.4.

Company Practices

Most companies have documented practices and procedures that describe how the company applies regulations, standards, and lessons learned. Additionally, large companies have their own internal guidelines for engineering and process safety.

5.3.2 Hazard identification and Risk Analysis

The *Hazard Identification and Risk Analysis* approaches described in Chapter 4 also apply to onshore production. Refer to Chapter 4 for further details. The main methods used include What-If, the What-If Checklist, and HAZOP (Hazard and Operability Study).

What-If

What-If examines operations for possible deviations to see what the consequences might be and what safeguards are employed. The safeguards listed may be barriers, barrier elements or degradation controls as discussed in the bow tie methodology (CCPS, 2018c). What-If or the What-If Checklist can be employed early in a process design (see Chapter 7 for further details). Teams review these for adequacy and may recommend additional safeguards.

Hazard and Operability Study (HAZOP)

HAZOP is a frequently used hazard identification method described in CCPS (2008a). A multidisciplinary team is convened to conduct the study. The process drawing is divided into nodes. A node might be a pipe flowing between two vessels, or the vessel itself. There can be multiple equipment items in a node including valves, pumps, filters, etc. A HAZOP study uses a selection of parameters (e.g., flow, level, temperature, pressure) combined with guidewords (no, more, less, as well as, part of, reverse, other than) applied to each node. This results in a rigorous identification of potential deviations. When a deviation is identified, the team assesses the potential consequences and documents any existing safeguards.

Risk ranking approaches are commonly employed to assist decision making. If the team thinks the safeguards listed are not sufficient, they may recommend additional ones. Further details on the HAZOP method and risk ranking are provided in CCPS (2008a).

Layer of Protection Analysis (LOPA)

The LOPA method is an order-of-magnitude semi-quantitative risk analysis technique. It is being increasingly employed in onshore production facilities to better understand the expected frequency of a specified consequence of interest. CCPS has published the method and two additional guidelines with suggested IPLs and conditional modifiers to use in LOPA assessments (CCPS, 2001, 2013a, 2015).

It should be recognized that the LOPA approach is not an identification method – it relies on HAZOP or PHA to generate the scenarios for further evaluation. Also, it considers one risk at a time and does not evaluate cumulative risks. This technique is frequently used during the design stage and is described in the CCPS references above.

Facility Siting

A process safety methodology often employed at larger onshore production sites is facility siting. This method is defined in API RP 752 for permanent occupied buildings and in RP 753 for temporary occupied buildings. CCPS (2018b) extends the API documents with additional guidance.

The basic aim of facility siting is to assess realistic leak scenarios and their predicted potential fire, explosion, and toxic outcomes. These impacts are overlaid on the facility plot plan and this is used to determine if there is sufficient spacing between the hazard source and occupied building locations. Further analysis can assess the layout of process equipment within a process unit. Consideration is given to barriers or barrier elements such as gas detection and ESD, fire and blast walls, drainage arrangements, etc., as these affect the predicted consequences.

An important aspect of facility siting is to identify what is considered an occupied space. This is not always obvious, and it is usually necessary to engage with personnel to identify these locations. The company must implement a system to ensure that, over time, unoccupied spaces do not become occupied as that invalidates the safe spacing decisions.

Vapor Cloud Explosion – Short Primer

Leak of hydrocarbon vapors or mists of flammable liquids disperse downwind. If this vapor cloud is ignited in uncongested space, then a flash fire event results. This is primarily a hazard to anyone trapped within the cloud. People close by wearing normal PPE, but not in the cloud, may not be seriously impacted. A much worse outcome occurs if the cloud disperses into congested space and a vapor cloud explosion results.

Process Safety Issues: The Flixborough chemical plant incident in 1976 was the first well documented VCE event and it, plus some serious offshore explosions such as Piper Alpha, led to much experimental work to understand the mechanism involved. Hundreds of large-scale experiments were carried out, mainly at Spadeadam in the UK and in Texas in the US. These showed that congestion from process equipment causes the flame front to accelerate from low speeds (under 10 m/s [33 ft/s] as in flash fires) to higher speeds (300 m/s [984 ft/s]) that result in damaging overpressures. This event is termed a deflagration. Deflagrations only generate overpressures for that part of a flammable cloud within the congestion, so the whole mass released may not contribute to the blast propagation.

It had been argued that some major explosion events were detonations. This was not initially accepted until detailed investigation of the Buncefield storage tank event in 2005 showed that it was a DDT event – deflagration to detonation transition. That event was caused by a tank overflow, spilling gasoline over a wing girder of the tank, and forming a mist of gasoline vapor and droplets at ground level. This was ignited and the flame front was accelerated, not by equipment congestion, but by dense foliage in a hedge row. A detonation event is more serious than a deflagration as the flame speeds are much higher (1000+ m/s [3281 ft/s]) and that all the material in the flammable range contributes to the explosion, not just the portion in congested space (Hansen and Johnson, 2015).

RBPS Application

Hazard Identification and Risk Analysis: Any team member in a HAZOP or Facility Siting Study can identify the potential for a vapor cloud explosion. Analyzing the severity of the explosion and estimating the risk of the explosion should be conducted by a technical expert in this field.

Barrier Analysis

The barrier approach as employed in bow tie analysis is now being used more frequently onshore. It is essential that personnel know what the important barriers against major incidents are and particularly those where they have a role in operating or maintaining them. Process safety events are rare and unless fully explained, personnel may not understand the role of barriers in preventing or mitigating these rare events. This contrasts with occupational safety events where the barriers are relatively easy to identify.

Knowledge of barriers also underpins effective implementation of MOC procedures. Barriers can be degraded during changes and such changes must be managed to ensure the barriers are returned to full effectiveness or are replaced with new ones.

5.3.3 Learning from Experience

As with well construction, onsite production facilities employ process safety programs to ensure the safety and environmental management system is fully functioning. Key RBPS Elements include *Incident Investigation, Measurement and Metrics, Management Review and Continual Improvement*, and *Auditing*.

To facilitate measurement and improving process safety performance, industry has developed standard process safety indicators as described in API 754 or OGP 456. Incidents should be recorded and categorized using these documents into one of four tiers. Many companies share this data (e.g., IOGP, 2019a), and this allows companies to benchmark their performance. Tiers 1 and 2 represent larger and medium size events. Tier 3 represents demands on safety systems (e.g., relief valve openings) and Tier 4 represents deficiencies in activities but no loss of primary containment (e.g., not meeting requirements of safety management system). Many companies have committed to publicly reporting the number of Tier 1 and 2 events, and some are also considering reporting Tier 3 events. These direct measures of process safety indicators help to identify areas for improvement.

Auditing is a formal review of all the management system elements to ensure that all specified processes are in place and functioning.

Management Review is a periodic ongoing process that assesses safety performance against current targets and determines whether existing controls are adequate or need improvement. If the aim is to drive performance to improve current targets, then that is termed *Continual Improvement*.

5.3.4 Emergency Management

Emergency Response Plans

Regulatory requirements for emergency response plans for larger onshore facilities are provided in OSHA PSM and EPA RMP or in the Seveso Directive. Smaller facilities follow state requirements or industry good practice. The RBPS element *Emergency Management* provides details. The emergency response plan should address both onsite personnel and offsite communities as well as external emergency responders.

The EPA (2009) has provided detailed guidance for the content of an emergency response plan. As well as various administrative requirements, for larger sites the plan must document a worst-case analysis for toxic and flammable cases, provide a 5-year incident history, and give details of the emergency response plan.

The RMP program is supported by an offsite consequence analysis (OCA) to provide a worst-case estimate for hazard distances due to explosion, fire, or toxic release. EPA (2009) provides guidance on how to conduct an OCA. The rules are detailed, but in simple terms the worst-case release is a total release of the largest inventory in the facility over a 10- or 60-minute period. As this gives very conservative results, companies usually also submit more realistic, alternative release scenarios. Endpoints for consequence distances are based on 1 psi for explosions, 5 kW/m^2 for thermal impacts, and ERPG-2 concentrations for most toxic materials. Manual calculation methods are provided; however, most companies use software (e.g., PHAST, Canary, Aloha, RAST). These models are for flat terrain, and if the local topography is not flat or has significant obstacles, then Computational Fluid Dynamics (CFD) software may be used to better represent the impact from topography and obstacles (e.g., FLACS, KFX). Local emergency responders use the calculated hazard zones of the OCA to plan their possible responses and required shelter in place warnings.

The emergency response plan should cover the following.

- Key contacts
- Alarm systems, CCTV and wind indicators
- Evacuation routes and alternatives
- Fire equipment and resources
- Mutual aid schemes
- Local emergency responders

Fire Protection

There are a wide range of fire protection measures employed at onshore production facilities. In bow tie terminology, fire protection barriers lie on the right hand or mitigation side of the diagram as they come into play after a loss of containment event. Normal elements of the fire protection system include the following.

- Fire and smoke detection
- Flammable and toxic gas detection
- Emergency shutdown systems (ESD)
- Dikes and drainage control
- Passive fire protection (PFP)
- Fire hydrants, monitors and water supply
- Water spray and deluge systems

- Sprinkler systems
- Fixed and semifixed foam systems
- Foam storage and application systems
- Portable fire extinguishers
- Special fixed extinguishing systems (e.g., FM200, INERGEN)

Additional guidance is available in CCPS (2003) and in NFPA and API standards. Fire systems involving logic controllers are often defined as safety instrumented systems and guidance for these is available in IEC 61508 and 61511. If not covered by these standards, the equipment should be independent of the plant distributed control system.

Fire Hazard Planning

Fire protection at small unmanned production sites is often modest compared to large facilities, matching their needs economically. Larger facilities may use fire codes to define their systems. Another approach is to use Fire Hazard Analysis rather than simply to apply code recommendations. FHA is a scenario-based approach where potential leak events are assessed, and their impacts calculated. Each scenario should be developed and tested to verify that the planned fire responses are adequate and tested through tabletop and field drills. Some issues addressed in an FHA include the following.

- Calculating firewater required to apply onto the fire and simultaneously to cool adjacent exposures effectively
- Where fire hose is required, determining if firefighters can deploy and connect all the hose required in a specified time
- Determining if the drainage is sufficient to control spills of burning hydrocarbons, etc.
- Determining if firewater monitors have the range to effectively reach the fire and if remote operation is appropriate

6

Application of Process Safety to Offshore Production

6.1 BACKGROUND

Offshore oil production, particularly in deepwater, has become a major part of upstream operations. Offshore is often divided into shallow water and deepwater production. Shallow water is usually considered to be anything under 1,000 ft (305 m). Deepwater is anything greater than this and the term "ultradeep water" for fields in greater than 5,000 ft (1,524 m) water depth is also used. In the Gulf of Mexico, many shallow water fields are small and are declining in production.

Shallow water structures in the Gulf are simple steel jackets. When similar designs were first employed in harsher environments of wind and wave, such as the North Sea, some initial failures (e.g., Sea Gem 1965) showed that greater strength was required. This either strengthened the steel structure or introduced novel designs such as gravity based concrete structures. Some designs improved process safety by separating accommodation from processing on bridge linked platforms.

As well construction technology advanced and large fields were discovered in deepwater; floating production facilities became necessary as fixed structures are impractical. Deepwater floating facilities have several possible designs, each with advantages and disadvantages (see Chapter 2 for examples). All the designs enable oil and gas production and have separation facilities. Export by pipeline is typical while FPSOs allow for storing oil in the hull for subsequent transfer to a shuttle tanker. A more recent development for gas fields is FLNG – Floating Liquefied Natural Gas facility, which includes a liquefaction plant as well as the usual treatment facilities. It also stores its LNG product onboard and transfers this periodically to LNG carriers for export. FLNG facilities add extra complexity to offshore facilities including cryogenic processing and greater inventories of hydrocarbons.

In principle, the process safety risks in deepwater facilities are often higher than for shallow water designs as the economics of deepwater mean the facilities are bigger and more costly to develop and the number of crew on board is greater. However, all facilities are subject to loss of containment incidents that can give rise to serious consequences. A number of historical incidents are discussed later in this section. One feature of offshore facilities important for process safety is that the workforce usually lives on board and thus off-shift personnel are potentially close to hazards. Onshore, off-shift personnel are at home or in accommodation modules located more remotely. On-shift personnel may be close to hazards onshore or offshore, but offshore they may be sited above or adjacent to process hazards, whereas this is less common onshore.

An example of a relatively simpler and lower congestion offshore platform is given in Figure 6-1. This platform in 237 ft (72 m) of water has two main decks with a helideck on the left and a flare tower to the right. The upper deck houses living quarters, offices, a galley and various storage and production equipment; and the lower deck houses various other rooms and equipment including the Motor Control Center. Most equipment is outside, not in enclosed modules.

An example of a more complex and congested offshore production facility was Piper Alpha, which was located in the North Sea in 474 ft (144 m) of water. The module layout is shown in Figure 6-2. This shows most equipment to be in modules with tight spacing in both the horizontal and vertical directions. In Piper Alpha's original design, the accommodation was sited as far as possible from major hazard areas, such as the wellhead and separation modules. However, a regulatory change was implemented by the Government to require gas to be recovered rather than flared. A gas compression area was added as a modification and this was close to the accommodation. This module was a major hazard due to large amounts of high-pressure gas being processed.

Offshore production usually starts with flow from a reservoir as a mixture of oil, gas, and water. The production facility may be directly over a single wellhead or may be centralized to process the feed from multiple wells and may use a system of gathering pipelines. Subsea systems are growing in complexity from simple mixing of streams to also include some processing. Topside separation facilities are similar to those shown in Figure 5-2 for onshore as the raw feed is similar. Typically, 1st and 2nd stage separators are used at high and low pressure to degas the oil and separate the oil and water. Gas is treated as necessary (e.g., removal of H_2S, dehydration) and compressed for export, used for power generation on board, or

**Figure 6-1. Example of shallow water facility Gulf of Mexico
BSEE Panel Investigation into West Delta Block incident in 2014**

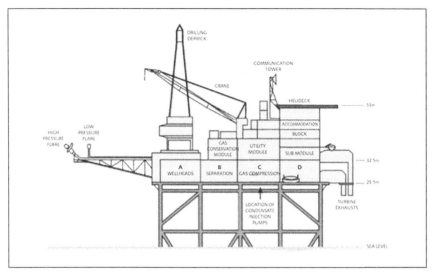

**Figure 6-2. Example of offshore layout
(Piper Alpha, McLeod and Richardson, 2018)**

reinjected if there is no export facility. Some gas treatment may occur onshore to simplify offshore processing. Similarly, oil is sometimes only partially stabilized with final stabilization carried out onshore.

One environmental operating factor of note for deepwater is cold sea temperatures near the wellhead, often around 4°C, even in warm climates. This can cause paraffins, waxy components, or hydrates to form as solids thereby creating plugging issues in the export line and risers. A process of flow assurance is used to keep the oil flowing. This may include injecting solvents or other inhibitors, or some combination of heated or insulated lines.

Historical Incidents

There have been several significant offshore incidents, and these have had a substantial impact on facility designs, safety and environmental management, and operational discipline as the industry responded. Two early events in the US affected designs and led to the stoppage of well construction off the coast of California. These were the Santa Barbara well control incident in 1969 and the Bay Marchand well control incident off Louisiana in 1970. Details of these events are readily available online. The Santa Barbara incident created a large oil spill zone affecting coastal locations and wildlife. There was a serious public outcry and the incident became a trigger for subsequent major environmental legislation. The Bay Marchand incident in 1970 caused four fatalities, lasted for three months, and created a large oil slick. In the North Sea, the Piper Alpha disaster in 1988 led to the far-reaching Cullen inquiry (available on the UK HSE website) and many changes to safety measures for offshore installations. The P-36 disaster in 2001 off Brazil highlighted additional safety issues and the need for operations excellence around maintenance isolation. The Piper Alpha and P-36 incidents are discussed later in this chapter.

Incident: Piper Alpha Disaster, July 1988

Piper Alpha was a major facility in the North Sea, handling both its own production and production from three other facilities. Initially, the facility only exported crude oil, but a later change in UK policy required that gas be collected and sent to shore as well. The platform was modified to include gas compression. However, blast walls were not identified as a required supplement to the existing fire walls.

In July 1988, a maintenance job was partially completed. It consisted of two separate tasks with two separate work permits. One job was finished, and the equipment secured, but the second (in an adjacent area) was not completed or secured. The status of the work permits at shift handover was not made clear. Operators started the unit, including the equipment where the work was not complete and the equipment was not gas tight, and a release immediately occurred. Gas started to accumulate, sounding gas detection alarms, but the gas cloud quickly ignited and blew out a firewall. The event escalated to other areas through the broken firewall. The control room was damaged and soon abandoned, leaving no means to continue to monitor or control the event. The Piper Alpha platform had been isolated by operators, but inflows from other platforms continued as their operators, who were aware of a problem on Piper Alpha, did not feel they had the authority to shut down their production.

About 25 minutes after the initial event, a main gas line ruptured, discharging 16-33 tons/sec (15-31 tonnes/sec) into the fire. The fire prevented the personnel onboard from reaching the lifeboats and they retreated into the accommodation area because it was fireproof. However, it was not smoke proof. No effective emergency evacuation was organized, and most personnel perished due to smoke inhalation. In total, 167 souls were lost.

A major public inquiry was held, chaired by Lord Cullen, which made many far-reaching, industry-changing recommendations regarding safe design, operation and regulation.

Source: Piper 25 Conference, https://oilandgasuk.co.uk › piper-25-conference

RBPS Application

Hazard Identification and Risk Analysis: A HIRA would have identified design changes that should have been made to reduce the risk from the new gas compression unit.

Safe Work Practices: The defective work permit system failed to communicate the status of the work or the condition of the equipment across a shift change. This set the stage for this tragic incident. This is similar to the P-36 incident, described later, which recommended an operational excellence program to ensure that safe work practices are always followed.

Emergency Management: Egress and evacuation plans were insufficient. There was no safe place for refuge and organization for safe evacuation.

The 100 Largest Losses in the Hydrocarbon Industry (Marsh, 2020) lists several major process safety events in offshore production or well construction that have led to large insurance claims. Many also involved multiple fatalities. Upstream incidents (all offshore) account for 23 of the total list of 100 major (financial) losses applying also to refineries, petrochemical plants, gas processing and terminals.

While there have been serious incidents offshore, industry data collection initiatives focusing on process safety as well as occupational safety are providing better transparency so that companies can better focus on process safety incidents – "what gets measured, gets done". The COS (2020) trend data for ten large companies and six contractors in the Gulf of Mexico for 2018 shows some improvement but also flat trends in some indicators. Specific Tier 1 and 2 process safety event statistics are available from IOGP (2019a) with both categories showing some good decline over the period 2011-2018 related to offshore activities.

6.2   OFFSHORE PRODUCTION FACILITIES: RISKS AND KEY PROCESS SAFETY MEASURES

Process safety risks on offshore production facilities are typically due to the inherent hazards of flammable reservoir fluids or other chemicals onboard and the activities carried out to support production. These include normal equipment operations and asset integrity activities, mechanical lifting, support vessel operations, mechanical degradation, etc. Improper management of these risks has the potential to cause a process safety event.

There are important aspects of offshore operations that increase risks compared to many onshore production facilities. These aspects include the following.

- The potential for fire or blast escalation given the limited footprint offshore that requires processing units/areas be located adjacent to one another or stacked vertically (as is apparent in Figure 6-2). The closer spacing creates a greater potential for even a small process safety event to escalate to a larger one by involving nearby process equipment. The facility design should account for these conditions through the use of *Hazard Identification and Risk Analysis* and consider emergency isolation or the addition of fire or blast walls to minimize the likelihood of escalation. The addition of walls can also trap hydrocarbon liquids and vapors, so the installation of these walls requires careful analysis and modeling. Weight restrictions offshore can limit the number and scale of fire and blast walls.
- Personnel often live onboard near process equipment, increasing their exposure to hazards and associated risks.
- Mechanical lifts are inherently more hazardous offshore due to lifting from different deck levels, supply vessels subject to sea conditions, and lift paths potentially over process equipment on deck and production infrastructure on the seabed. An advantage offshore, however, is that crane bases are fixed structurally, and hence toppling events are less likely than onshore.

- Helicopter landing and take-off operations and offshore vessel operations can potentially impact process equipment or risers through collisions.
- Process facilities are often contained within enclosures to protect equipment and workers from the weather and this increases the risk of small leaks accumulating to flammable concentrations that might otherwise disperse safely in an open design. It is worth noting that some onshore facilities, especially in harsher weather locations, are also contained within enclosures and share this risk.
- SIMOPS is a key challenge for offshore. Refer to Chapter 5 where this topic was discussed.

The following sections highlight some of the more common hazards associated with areas of an offshore facility. A *Hazard Identification and Risk Analysis (HIRA)* is recommended to determine the possible hazards and incident scenarios applicable to a specific offshore facility. Industry codes such as API RP 14C, 14J and ISO 17776 are helpful for hazard identification. The UK HSE provides a wide range of useful notes and regulatory guidance on offshore hazard management on its webpage (www.hse.gov.uk/).

6.2.1 The Well

Risks

Loss of well control can occur during the production phase in addition to during well construction. This may be due to interventions or workovers as an extension of the well operations or due to production problems or collisions. These were discussed in Section 4.2.2 summarizing the SINTEF blowout database.

Key Process Safety Measure(s)

Asset Integrity and Reliability: There are several asset integrity activities undertaken to ensure that the wellbore maintains its integrity during production operations.

Emergency Management: The BOP is replaced during production by a Christmas tree and this is closed to prevent a potential loss of containment. Subsurface safety valves (SSSV) are often installed into the well as an additional barrier. A loss of well control does not automatically mean a loss of containment to the environment. If one barrier is lost, the well should be shut in until actions are implemented to stabilize the well and restore the lost barrier.

6.2.2 The Production and Export Risers

Risks

The production riser takes production from wells on the seabed or from nearby facilities up to the topside production facility. It also takes cabling and other services down to the wellhead or other facilities. Export risers send oil or gas down to export

---

Incident: P-36 Incident, March 15, 2001

The P-36 FPU was located off Brazil, producing oil and gas in deep water. A design safety concept was for all hydrocarbons to be located on the topsides, with none located in the four vertical columns or the two submerged pontoons. However, drain tanks were located in two of the columns to receive separated water before discharge. One of these tanks was taken out of service for extended maintenance and all the tank connections were isolated correctly using blind flanges, except one which used only a block valve for isolation.

On March 15, a loud "bang" was noted in the column and an emergency response team was sent to investigate. As the column had been designated a "safe place", no special precautions were taken by the team. When they started their entry, a flammable cloud in the column ignited and 11 crew members were killed. The blast also sheared the main cooling water line and the rig started to flood. Since the cooling water also supplied the firewater system, it was by design not easy to isolate the incoming water. The rig was evacuated and after five days, even with salvage efforts, the FPU eventually sank.

Investigation showed that, over a period of time, the drain tank block valve leaked slowly, and a mixture of water and some oil flowed into the isolated drain tank. As the relief valve had also been isolated, the internal pressure rose and eventually a pressure burst occurred – this was the loud bang. The ruptured tank released flammable vapor that was ignited by the entering emergency team. The primary result of the Petrobras investigation was a recommendation to implement a system of Operational Excellence to ensure that all maintenance work including isolations was conducted correctly. Several technical improvements were also implemented across Petrobras.

Reference: Barusco (2002) The Accident of P-36 FPS

RBPS Application

Safe Work Practices: Maintenance work isolations were not performed correctly. Had the drain tank been fully isolated with blind flanges, then the incident would not have occurred.

Process Knowledge Management: The response team did not understand the potential for a designated "safe place" to become unsafe as there was a process vessel in the leg. Emergency teams should carry out gas testing before entering confined spaces.

---

pipelines on the seabed. Production and export risers are susceptible to collision by vessels (such as supply vessels or crude shuttle tankers in the case of an FPSO) or to dropped objects. Either can cause a riser leak or rupture and can result in environmental pollution. If the hydrocarbons reach the surface and ignite, a fire on the sea surface occurs that can impact the facility. An example of a riser rupture due to collision is the Mumbai High incident (refer to Table 1-1). A means to protect against riser collision is by locating risers on the inner side of the platform legs.

A special risk for floating production facilities (e.g., FPSOs and FPUs) is drift-off if the fixed mooring system fails in heavy weather. This can rupture the riser. The design includes auto-disconnect features to prevent a rupture event. US regulations require that offshore installations shut down and evacuate during certain hurricane conditions, thus reducing the risk of a loss of containment event.

An additional risk is fatigue failure of the riser connection to the wellhead due to vessel movement or vortex induced vibration from ocean currents.

Key Process Safety Measure(s)

Hazard Identification and Risk Analysis in the initial design identifies threats and incorporates appropriate risk controls. For example, with fixed platforms, risers can be placed inside jacket legs, riser guards can be added, or other physical protection installed. Isolation valves can be installed (e.g., a Christmas tree on the topside and SSSV set below the mudline) limiting any loss of containment to the content of the riser. Export risers can be protected using topside ESD systems preventing export oil or gas from entering the riser, as well as bottom isolation valves to prevent backflow from the export pipeline.

Riser fatigue failure is assessed by detailed assessment of ocean currents and potential vessel movement. The design should accommodate these forces.

Asset Integrity and Reliability programs assess the condition of the risers and determine if there is any degradation due to fatigue or corrosion so that this can be rectified before any failure occurs.

Safe Work Practices are employed to minimize the potential for dropped objects or collisions with supply vessels. This is discussed in Section 6.2.5.

6.2.3   Topside Production Equipment

Risks

There are many potential causes for loss of containment of well fluids or other chemicals from production equipment (e.g., corrosion/erosion, dropped object, operational upsets or maintenance issue). Potential topside production equipment leaks offshore are broadly similar to the risks for onshore facilities listed in Section 5.2.1 since the processes are similar. However, the consequences can be more severe due to tighter spacing, congestion and confinement and the closer proximity of personnel workspaces and accommodation.

Some of the primary potential sources of hydrocarbon leaks topside include the wellhead, the separation vessels (1st and 2nd stage and the test separator), the export pumps, plus all the associated pipework. Gas leaks could also arise in the wellhead, separators and from the various gas treatment equipment. The gas compression system is a large source of high-pressure gas. The power utilities, which also usually run on gas, are a potential leak source.

Once containment is lost, hydrocarbons accumulate, initially near the source location and then in other locations if not controlled. If ignition occurs, the event can escalate to adjacent modules (horizontally or vertically) if rapid isolation is not possible or if firewalls or blast walls cannot withstand the load. A short primer on the potential for flame acceleration and blast impacts to occur was provided in Section 5.3.2.

Key Process Safety Measure(s)

Hazard Identification and Risk Analysis: During engineering design, HIRA identifies prevention and mitigation measures to manage various loss of containment scenarios. Prevention controls normally relate to good operations, and maintenance and inspection activities. Mitigation measures include ventilation systems to dilute or extract smaller leaks, ignition controls, fire and gas detection system, emergency shutdown system, a depressurization and blowdown system to de-inventory the affected area, and a drainage system. Fire and blast walls, if installed, mitigate the potential for escalation. A backup battery power supply system provides power for some period of time to the control room and key facilities if power is lost. The firewater system usually has at least one diesel powered fire pump that operates without electrical power. The complex alternatives and rapidity of event progression tend to encourage at least some automated response systems. The need for some or all these barriers would be determined by *Hazard Identification and Risk Analysis*.

Emergency Management: If ignition occurs, the active and passive firefighting system comes into effect, and personnel follow evacuation, escape and rescue procedures (see Section 6.3.5) to reach a safe refuge or evacuate the facility.

6.2.4 Oil Storage Tanks

Risks

Most offshore facilities export oil by pipeline once it passes through the separation system. However, some designs include local oil storage, such as with FPSOs, which store oil in large tanks in the body of the vessel prior to periodic offloading by tanker.

The risks relate to spills of large volumes of oil from a tank if it is punctured due to a collision or major pipe rupture, and subsequent fire on the sea surface or environmental pollution.

Additional storage of hydrocarbons offshore may include helicopter fuel and diesel fuel for power generators. Leaks from these tanks can cause a process safety event.

Key Process Safety Measure(s)

Compliance with Standards: FPSOs are ship-shaped facilities and are covered by marine classification requirements (e.g., from ABS, DNV GL, or Lloyd's Register) if they are capable of self-propulsion, even if they are permanently moored. This is an International Maritime Organization rule, enforced by maritime regulators globally. Classification rules have detailed requirements for safe storage of oil in onboard tanks and environmental protection, like those for oil tankers. The rules are prescriptive and focus on design requirements. Periodic surveys are required to verify that the facilities remain fit for duty.

While maritime regulations require double bottoms, they do not require double sides, so a puncture to a single side wall design can result in a potential spill and environmental impact. Explosions are unlikely, but if air enters the tanks, for example when oil is being exported, then a flammable mixture in the vapor space can exist. Inert gas generators are used to fill the space with oxygen free vapor.

6.2.5 Other Offshore Risks

Additional process safety risks exist and are identified below.

Dropped Objects

Dropped objects can be a source of process safety incidents where these impact piping, vessels, wellhead, the risers, and equipment on the seabed. Offshore facilities are mostly supplied by marine vessels. Cranes are most frequently used to transfer loads from the vessel onto the facility, which may require lifts up to and exceeding 50 m (165 ft), and sometimes require transfers over process equipment. Once on the facility additional lifts may move the object to its final location using winches, forklifts, or rig hoisting equipment. The combination of wave motion on the vessel, currents, as well as wind forces or the possibility of rigging and crane failures, makes dropped objects a credible risk.

Dropped objects on the topsides are relatively straightforward to evaluate; however, objects falling into the sea may not travel straight down due to their shape, and process equipment on the seabed not directly below the fall location is at risk of impact damage.

Collisions

Offshore installations are serviced by supply vessels, personnel transfer vessels, and potentially by shuttle tankers for offloading. Also, installations near sea lanes are passed by many commercial vessels. These can all collide with the offshore installation and lead to leaks of process fluids (e.g., supply vessel collision with Mumbai High in 2005). Similarly, helicopters are used for personnel transfers and these can collide with process equipment. While these events are not common, a full hazard identification evaluates the potential and whether key lines should be protected from possible impacts.

Vents and Flares

The risks of flammable or toxic vapors are the same as discussed for onshore production in Section 5.2.6. An issue for offshore is that the footprint is smaller and the flare is usually located on a flare boom cantilevered to one side of the facility. This must be away from the accommodation or the helideck. The design should ensure that thermal radiation levels meet API 521 guidance. However, since spacing is tight, predicted radiation may be close to the limits rather than a lower level, which more generous spacing onshore might allow.

Vents have an additional risk for offshore floating installations. In an emergency where buoyancy is partially lost and the offshore facility lists, vents become an entryway for seawater into the hull. This can exacerbate the buoyancy problem and lead to sinking and a consequent process safety loss of containment incident. Discharge and inlet vents are often grouped near one another, although this can have undesirable consequences. In the P-36 incident, the initial gas alarms were confusing as vents discharged flammable vapors into the atmosphere as planned, but nearby air inlet vents sucked this down into other modules, setting off their flammable gas alarms. This made diagnosing the source of the flammable vapors confusing.

Subsea Currents and Loop Currents

Subsea currents, also known as a loop currents in the GOM, can cause subsea pipelines or risers to move or cause fatigue failures and hence potentially lead to a loss of containment incident. The inherently safer approach suggests it is best to address these risks during design. Remaining risk can be addressed through other controls such as defining operating windows.

Key Process Safety Measure(s)

Hazard Identification and Risk Analysis: Studies identify the dropped object risks and inform layout decisions to provide adequate laydown areas for loads and adequate corridor space for forklifts, etc. These findings cascade into specific *Safe Work Practices, Conduct of Operations*, and *Training and Performance Assurance* elements of RBPS. Design studies predict vent discharge concentrations, and these are used to avoid interaction between discharge and inlet vents.

Process Knowledge Management: Understanding complex subsea currents is needed to evaluate risks to subsea pipelines, risers, and other equipment.

Safe Work Practices: The avoidance of collisions and dropped objects is managed by safe work practices and procedures.

Emergency management: Various isolation and shutdown systems are used to isolate a leak (as discussed in Section 6.2.1) caused by a dropped object or collision or leaks due to subsea currents.

6.3   APPLYING PROCESS SAFETY METHODS IN OFFSHORE PRODUCTION

Most of the process safety methods and tools discussed previously in Chapters 4 and 5 also apply offshore. Where this is the case, only key differences are discussed in this chapter. LOPA is a commonly used technique during design for both onshore and offshore production facilities.

6.3.1 Process Safety Culture

Process Safety Culture applies to all life cycle stages for onshore and offshore. Refer to Section 4.3.7 well construction where this topic was discussed. A positive safety culture is key to achieving exemplary offshore process safety.

The Presidential Panel investigating the Deepwater Horizon incident identified several aspects of deficient *process safety culture*, even in the companies that had excellent occupational safety performance. The National Academies (2016), following on from the Deepwater Horizon incident, reviewed the need for an improvement to safety culture for offshore operations to prevent major incidents – this is what CCPS terms *process safety culture.*

6.3.2   Regulations and Standards

An overview of regulations was provided in Section 2.8. As with onshore production (Chapter 5), offshore production is developed based on a collection of regulations, codes, standards, and industry practices.

Regulations

In the US, the US Coast Guard (USCG) and BSEE split responsibility for offshore safety – with the USCG focusing on maritime, structural and evacuation operations, and BSEE addressing well construction and production operations. In the US, prescriptive regulations address many aspects of offshore design, often by reference to standards from API or using requirements or guidance from COS. SEMS regulations were a departure from the prescriptive approach for BSEE and are described as goal-based. Floating installations often follow marine and offshore classification rules issued by classification societies, which are also mostly prescriptive. Many international regulators use goal-based safety regulations (EU regulators, UK HSE, Norway PSA, Australia NOPSEMA, Canada C-NLOPB), underpinned by a thorough risk analysis, to identify major risks and important safety barriers with the aim to reduce risks to ALARP. These regulations are normally paired with prescriptive requirements to give a blended form of regulation.

Standards and Guidelines

The API has issued many RPs addressing management systems, safety topics and technical issues for offshore operations (see API, 2015a). Offshore RPs are similar to onshore RPs, but address issues unique to offshore facilities and operations. A study by ABSG (2014) provides a useful comparison of API and international offshore safety-related standards and how these relate to BSEE SEMS regulations. Other standards from ISO, IEC, EU and NORSOK are also widely used in offshore production. Norwegian standards NORSOK S-001 (2018) on Technical Safety and Z-013 (2010) on Risk and Emergency Preparedness Assessment provide detailed guidance for safety studies. IEC 61511 and ANSI / ISA 84 provide guidance on implementing safety instrumented systems. In the UK, FABIG (Fire and Blast Information Group, a part of the Steel Construction Institute) provides a useful series of papers and conferences on offshore safety technologies.

Company Practices

Internal company practice documents incorporate current good practice and lessons learned from their operating experience on how to manage their risks. These cover guidance on management systems, risk management, asset integrity, and interpretation of technical standards when those standards offer multiple options. There is no single style to company practice documents; some are prescriptive, and others are goal-based, starting with a risk assessment. RBPS (2007), API RP 75 4th

Ed (2019a), COS (2014) and UK HSE HSG 64 (2013) all provide practical guidance on leadership roles and what should be covered in safety management systems and their documentation.

6.3.3 Hazard Identification and Risk Analysis

Inherent Safety

The RBPS element *Hazard Identification and Risk Analysis* is applied to offshore platform design with the aim to enhance inherent safety (discussed in greater detail in Chapter 7). Generally, this is similar as for onshore facilities and, in addition, recognizes special issues of tight spacing, potential escalation, and personnel exposure. It starts with identification of hazards, development of designs or revision of designs to mitigate some of these hazards, and provision of adequate barriers to control the remaining hazards. Some useful process safety themes contributing to inherent safety might be as follows.

- Developing facilities to be normally unmanned
- Reducing inventory – smaller vessel dimensions
- Less processing offshore - with more processing onshore
- Power generation onshore rather than onboard (as in Norway)
- Enhancing communications to easily engage with process safety and facility engineers, and other support onshore
- Designing for well fluids that change over time
- Separating occupied areas (accommodation, control room) from higher hazard areas (e.g., wellheads, separation modules, compression modules) and using bridge linked platforms for accommodation when practicable
- Designing modules to allow for adequate operations and maintenance access and any necessary lift activities
- Reducing potential blast propagation (e.g., more open space when possible and using safety gaps on large FLNG facilities)

Internationally, many countries follow a safety case approach as was described in Section 2.8. The safety case is based on a detailed risk assessment (RBPS element *Hazard Identification and Risk Analysis*) and demonstrates that feasible risk reduction has been achieved, that suitable risk management and safety barriers are in place, and that emergency response and environmental protection is well planned. Risk management follows the full set of RBPS elements (Chapter 3).

Hazard Identification and Risk Analysis Tools

A number of hazard identification and risk tools may be appropriate to analyze a design and provide for escape, evacuation and rescue. These may include, amongst others: fire and explosion analysis, analysis of occupied spaces, gas dispersion analysis (as relates to air intakes and helidecks), adequacy of temporary refuge studies (fire, smoke, and toxic gas), blowdown and pressure relief, and emergency systems survivability analysis. Some companies also develop voluntary safety cases even if not required by local regulations.

Hazard Identification

The HIRA tools used here are similar to those discussed earlier – What-If, What-If Checklist, HAZOP and qualitative or quantitative risk assessment. There are specific industry guides that assist offshore hazard identification, rather than starting the exercise from first principles only.

A basic set of design guidance is found in API RP 14C (2017). It contains multiple design safety features for all parts of the offshore production facility. The approach does not focus on identification of hazards; rather it provides suggested designs to enhance safety of that part of the facility. By checking designs against this reference, this provides an indirect form of hazard identification as any omissions of safety equipment can be investigated further.

An international standard ISO 17776 (2016) provides greater detail for the hazard identification process. The standard provides a hierarchy of identification methods from judgement to fully structured methods. No one method is always suitable for all facilities, so the operator should consider the complexity of the facility and the inherent hazards (hazardous inventories, people at risk, etc.) to make an appropriate selection. Once hazards are identified then risk evaluation is recommended to determine if additional risk reduction measures are required. A schematic for these approaches is shown in Figure 6-3. The standard provides a suggested risk ranking matrix that has been found to work effectively (reproduced earlier in Figure 4-4).

ISO 17776 introduces concepts for barrier-based thinking for offshore. It provides multiple tables with lists of hazards for all phases of offshore operations. It also breaks out details for each hazard category and uses a standard numbering system which assists the team to check that they have addressed all hazards and for others to verify the quality of the assessment.

An additional, more detailed standard for hazard identification, is API RP 14J (2001). This is a hazard identification guide that provides useful background on a wide range of hazards in all units of an offshore facility. The 14J RP reviews several identification methods, including checklists, What-If analysis, HAZOP, FMEA, and Fault Tree Analysis. Some example checklists are provided. A discussion is provided on good layout principles – mainly separating the higher hazards from occupied areas. It refers readers to CCPS Hazard Evaluation Procedures Guide (CCPS, 2008a) for details on methodology selection and execution.

Fire and Blast Studies

Fire and blast studies are a part of the RBPS element *Hazard Identification and Risk Analysis*, and usually follow from potential scenarios identified in HIRA studies. It is the aim of fire and blast studies to ensure sufficient barriers protect the facility from escalation and to assure that refuges and evacuation areas are protected from these events.

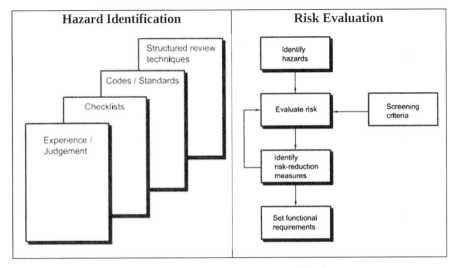

Figure 6-3. Hazard identification and risk evaluation approaches (ISO 17776, 2016)

Fire studies are similar to those covered in Chapter 5 for onshore production facilities. API RP 2001 (2012) for refineries provides useful guidance, some of which is applicable for upstream (onshore and offshore). The UK HSE (2009) provides guidance for a scenario-based fire protection assessment. FABIG (2014) provides useful guidance for fire protection designs for offshore installations. Generally, a scenario-based approach is used as few API or NFPA fire related standards directly address upstream production facilities. Some important differences for upstream relate to, for example: the layout offshore has closer spacing and vertical stacking of units not common onshore leading to greater escalation risks; hydrants offshore have obstructed views, particularly where equipment is in modules, making deluge and sprinkler systems more important for cooling; and escape and evacuation is more difficult.

As mentioned previously, escalation is one of the more significant concerns for offshore facilities. For scenarios that are not sufficiently mitigated by fire and gas detection and ESD systems, fire and blast walls are two important additional mitigations. Fire walls are designed to withstand specific types of fire (cellulosic, hydrocarbon pool, or pressurized jet) for a defined period. For example, an H-60 firewall is designed to withstand a hydrocarbon pool fire for 60 minutes.

Blast walls are used to separate equipment vulnerable to escalation due to vapor cloud explosion. This might be due to large inventories (e.g., separation module), operations at higher pressures (e.g., the compression module), or to protect significant personnel exposures (e.g., in accommodation or control room). NORSOK Z-013 (2010) provides risk-based guidance on how to determine fire and blast loads (in the form of design accidental loads) and hence the required fire and blast wall specifications. A downside of fire and blast walls is that they contribute to possible confinement of a flammable cloud and this increases the consequence severity.

Quantitative Risk Assessment

Quantitative risk assessment (QRA) is part of the *Hazard Identification and Risk Analysis* element of RBPS. It is being used offshore, primarily in countries following the goal-based approach to safety regulation (e.g., UK, Norway, Australia) or where internal company practices suggest its use. The main purpose for carrying out a QRA is risk reduction and to assist with making complex decisions on the provision of safety equipment. It provides insights on cumulative risk not available from LOPA which examines risk one scenario at a time, and how much risk reduction is achieved for a given investment. Judgement alone cannot make such decisions in a rigorous manner. At the early design stage, a concept QRA may be used to select between different design alternatives (e.g., spar vs TLP vs FPSO). Once the final design is delivered, a full QRA can be used to assess the design changes made along the way and provide a final assessment of key risk measures. There is value in conducting a QRA for complex installations irrespective of regulatory requirements as it makes transparent design assumptions that warrant preservation during management of changes over the facility life cycle.

The methodology for an offshore QRA is similar to that discussed in CCPS (1999) for CPQRA (chemical process QRA), but with an explicit need to evaluate escalation as part of an event sequence and to understand the risks of impeded egress, escape and rescue strategies that are much more pronounced offshore. These events, termed domino events in downstream, are often not needed in downstream risk evaluations as spacing is sufficient to prevent escalation and provide ample egress and escape options. A Norwegian standard (NORSOK Z-013, 2010) provides detailed guidance for an offshore QRA, while a simpler approach is described by Spouge (1999). Vinnem and Roed (2020) describe several different QRA approaches as well as lessons relevant to QRA studies from a thorough review of 35 different offshore incidents.

Consequence software is usually employed as there are many outcomes to track for each leak source (i.e. different hole sizes, orientations, and ignition possibilities). This usually requires proprietary software, either integral models (e.g., PHAST, Canary) or more complex CFD tools (e.g., FLACS or KFX) where simpler models are not appropriate.

Frequency and probability data are unlikely to be available from a single company as major leaks are rare events, and many leaks that need to be considered are smaller than the Tier 1 and 2 thresholds (API RP 754 and IOGP 456) being tracked by many offshore companies. IOGP (2019b) has issued an update to its offshore process equipment leak frequency datasets, as well as other important data sets on riser and pipeline leaks, blowout frequencies, and ignition probabilities. This data was developed for IOGP by DNV GL from the UK HSE Hydrocarbon Release Database. IOGP (2019b) also issued data on blowout frequencies for use in QRA based mainly on the SINTEF blowout database and Lloyd's Register assessment of that data.

Additional factors to consider include ignition probabilities, personnel presence factors, escalation outcomes, etc. The risk contributions from all scenarios are summed and a location specific individual risk (LSIR) developed for all modules, as well as a combined group risk metric, the Potential Loss of Life (PLL). LSIR and PLL results are probed to identify the most hazardous locations and the major contributor events so that risk reduction measures can be targeted. Moreover, these measures are evaluated for the risk reduction achieved. Some companies use the ALARP principle for this determination. In simple terms, ALARP tests whether the cost of a measure in terms of time, money, and effectiveness is commensurate with the risk reduction achieved.

Software is available for performing an offshore QRA. A simple approximate spreadsheet method widely used for the UK Sector of the North Sea was developed by Spouge (1999). More detailed software tools are available which carry through the calculations and escalation potentials without the simplifications of spreadsheet methods (e.g., SAFETI Offshore described in Pickles and Bain, 2015). BP has developed its own screening offshore QRA tool, OMAR. These software tools generally do not include CFD directly, due to the computational burden, but summary CFD results can sometimes be imported. The tool used should be commensurate with the type of decision, data available in the project stage, and the type of result and precision required.

Fault and Event Tree Analysis

Fault trees are a form of risk analysis and are part of the RBPS element *Hazard Identification and Risk Analysis*. Fault trees are useful to establish causal links in an event sequence and are most commonly applied offshore to understand BOP failure modes and overall reliability.

The method is described in CCPS CPQRA (1999) and the Hazard Evaluation Procedures Guide (CCPS, 2008a), and in more detail in NASA (2002). Fault trees start with initiating or base events and show how a combination of events or barrier failures allow the initiating event to progress upwards to the next step, and ultimately to the top event (usually a loss of containment or system failure). Event combinations are achieved using AND or OR gate logic. AND logic requires both inputs to be true to progress upwards, while an OR gate allows the progression if any one event is true. An example is shown in Figure 6-4. In this figure, boxes are events, and boxes with circles below are base events. AND gates are represented as an arch with a flat bottom and OR gates as an arrowhead shape.

Fault trees help identify common mode failures, where a single failure disables multiple barriers and greatly increases the risk in a manner that can be hard to visualize without this analysis. Another FTA application, termed minimal cut set analysis, helps identify the fewest number of failures that lead to the undesired top event occurring. Single event cut sets are very serious as a single failure results in the top event.

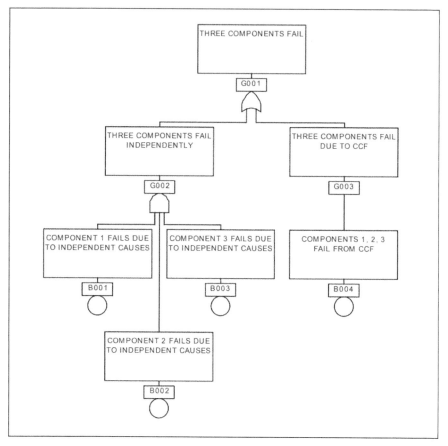

**Figure 6-4. Example fault tree logic
(from NASA, 2002)**

Another form of *HIRA* analysis is Event Tree Analysis, also described in *Guidelines for Chemical Process Quantitative Risk Analysis* (CCPS, 1999). Fault trees build to the top event. The event tree takes this event through the many possible outcomes depending on whether safety barriers are effective. An example of the many possible outcomes is shown in Figure 5-2.

6.3.4 Asset Integrity and Reliability

Topics related to safety systems are addressed in a number of RBPS elements. It is convenient to address them under *Asset Integrity and Reliability* given the importance of safety system reliability.

Safety Critical Systems and Equipment

The term 'Safety Critical Elements' is used in the UK to describe those controls put in place to prevent or mitigate significant process safety events. These may be full barriers or individual barrier elements (see Section 2.7). These must be identified and subject to regular inspections, testing and maintenance programs to ensure their operability and reliability. Safety critical elements are not specified by specific codes and standards. They must be identified based on the specific process safety risks unique to a given operation and the risk acceptance criteria used.

Safety critical elements are devices, machinery, instrumentation, controls, or systems whose malfunction or failure likely results in a release of hazardous materials or energy with major consequences such as toxic exposures, fires, explosions, property damage, lost production, environmental damage, and fatalities. The identification of safety critical systems or equipment is part of the overall process safety and environmental management system and is addressed in the RBPS element *Hazard Identification and Risk Analysis*. The process is described in CCPS (2016) *Guidelines for Asset Integrity Management* and EI (2007) *Guidelines for the Management of Safety Critical Elements*.

One example of a set of safety critical elements defined by an operating company is provided in Table 6-1. While this is a specific example, other facilities are likely to be similar. The elements in this table are higher-level "systems", i.e., above the individual part level. For example, "Active fire-fighting systems" potentially includes one or more fire water pumps, a fire water ring main, hydrants, monitors, deluge nozzles, valves, etc. Some of these individual pieces of equipment significantly affect the performance of the system and should be considered safety critical as well (i.e., at the equipment tag level).

Table 6-1. Example safety critical systems and equipment

No.	Description	No.	Description
1	Containment design	11	Emergency power and lighting
2	Natural ventilation and HVAC	12	Process safety equipment
3	Gas detection	13	PA, alarm and emergency communications
4	Emergency shutdown system	14	Escape routes and evacuation
5	Open drain system	15	Explosion barriers
6	Ignition source control	16	Cranes
7	Fire detection system	17	Drilling and well intervention equipment
8	Blowdown and flare/vent system	18	Ballast water and position keeping
9	Active fire-fighting system	19	Ship collision avoidance systems
10	Passive fire protection		

Safety critical elements demonstrate four objectives: functionality (does it achieve the desired result), reliability (does it have an acceptably low probability of failure on demand), availability (is it functional for an acceptable percentage of time when the facility is expected to be operating), and survivability (the element's ability to survive the event for which it must respond). The Pryor Trust incident described in Chapter 5 shows the importance of survivability. The drilling rig should have been designed and maintained such that the hydraulic control lines actuating the BOP remained usable for an adequate period of time following a blowout-induced fire so that the BOP could be closed. In fact, the control lines failed quickly and the BOP could not be closed.

6.3.5 Emergency Management

Emergency Management is an element of RBPS as well as being included in API RP 75 and in safety case requirements. Generally, this element comes into play after there is a loss of containment event, but some aspects are designed to minimize escalation of the incident and others to protect personnel and the asset. This section considers the fire and gas detection system and fire protection equipment as examples in SEMS and the escape routes and evacuation and rescue plan as an example of the safety case regulations. Testing and drills are critical to maintaining the effectiveness of the system.

Fire and Gas Detection System

As mentioned previously, *HIRA* studies underpin almost all of RBPS, including *Emergency Management* planning. Hazard identification establishes the need for fire and gas detection and risk analysis can be used to establish the number, location, and voting logic (if any) for the detectors. Fire and gas systems that detect a fire and initiate a shutdown are usually designed as a SIS (safety instrumented system) following guidelines such as IEC 61511 or national equivalents (e.g. ANSI / ISA 84). SIS systems have three elements – detectors, logic solver and actuators, which as an instrumented function must deliver the specified reliability. The logic solver is usually a computational module and the actuators are fast acting emergency block valves. A human can take the place of a safety instrumented function if they meet the criteria required to determine the problem (detect), understand what should be done (decide), and execute the task at hand (deflect), such as manually activating an ESD system.

Flammable gas detectors along with flame/smoke detectors form the detection part of a fire and gas system. The number and location of detectors can be aided by historical rule sets for spacing, CFD-based scenario modeling, and by other approaches. Scenario analysis from the *HIRA* identifies the lower bound size of gas leaks that should be identified as no system can alarm all leak events quickly.

Basic approaches use simpler consequence modeling tools to predict flammable envelopes, and there is increasing use of CFD tools to better account for gas flows around obstacles common in offshore modules.

Fire protection

Firefighting capability (water and foam) is available from hydrants, sprinkler and deluge systems on modern offshore installations and from monitors on offshore support vessels. One advantage offshore is there is an abundance of firewater supply – so long as power is available. However, there is limited line of sight to attack fires inside modules or deep within a congested layout other than using sprinklers or deluge systems installed above vulnerable equipment.

Fire protection is a mix of active and passive systems. Active systems include fire and gas detection, alarms, fire water pumps, fire water ring main(s), hydrants/monitors, water sprays, foam systems, and deluge systems. Emergency shutdown and isolation of hydrocarbon flows is also an important element of active fire protection. Passive systems include fireproof coatings, fire walls, blast walls and drainage.

Escape, Evacuation and Rescue

An aspect for offshore operation is the greater difficulty of personnel to escape compared with onshore facilities. Jumping into the water is typically not a good option due to height, ocean conditions, and oftentimes remoteness from shore or other facilities. Offshore facilities typically conduct an EER (Escape, Evacuation and Rescue) study to plan for significant process safety events and to make sure escape pathways and safe refuges are defined, and adequate equipment (e.g., lifeboats) and plans are in place. Temporary safe refuges (TSRs), or temporary refuges, are designed to protect personnel against the possible process safety event consequences for some defined period of time, especially fire or smoke, until an evacuation is ordered, or the situation is resolved.

Evacuation offshore is usually achieved using lifeboats, although there are supplementary systems using escape chutes and life rafts. The authority having jurisdiction may set minimum requirements for lifeboats / life rafts. Helicopter evacuation is an option but it may not be possible to land on the helideck if there are flammable vapors present, fire and smoke, or the offshore installation is listing. An inherently safer evacuation exists for bridge linked platforms, where evacuation is by walking over the bridge.

Rescue is typically achieved using an offshore support vessel, most of which have firefighting capabilities. These type of support vessels are mandatory in many offshore locations.

Oil spill response is another important consideration for offshore *Emergency Management*. Drain systems are designed to intercept hydrocarbon leaks preventing them from flowing overboard. Booms and other mitigation equipment are usually present (on the facility or on a nearby support vessel) to contain spills. After the Deepwater Horizon incident, operators in the US and many around the globe joined to form response consortia. These maintain dispersants, oil spill containment, and recovery equipment suitable for rapid deployment to any significant spill incident.

7

Application of Process Safety to Engineering Design, Construction and Installation

7.1 BACKGROUND

7.1.1 RBPS and Project Engineering

The process safety risks of well construction, and onshore and offshore production were presented in earlier chapters. Given this background in the operating phase, the process safety risks to address in engineering design, construction and installation activities are covered here.

Process safety activities during engineering design are just as important as during operations, although the activities may be different. This chapter highlights where RBPS integrates into every stage of a facility's life cycle. A detailed table is provided in CCPS (2019b) showing the element by element linkages during projects. Of potentially greatest importance, is the ability to implement inherent safety during the early design stage when options are on paper, rather than after construction when the hardware is in place.

A useful reference relevant for this chapter is CCPS (2019b) *Guidelines for Integrating Process Safety into Engineering Projects*. This book addresses all engineering projects, whether upstream or downstream, large or small. It uses a project life cycle approach, "the series of phases that a project passes through from its initiation to its closure" PMBOK Glossary (PMI, 2013), which is used in this chapter. The Construction Industry Institute at the University of Texas has issued useful guidance as well.

While all of RBPS applies, the most important pillar and elements during engineering design might be as follows.

- *Pillar: Commit to Process Safety* - A Process Safety Design Philosophy may be developed and all the PS activities mentioned below should be listed in that document along with the timeline (project stage) and scope.
- *Pillar: Understand Hazards and Risks* – It is during design that a full understanding of hazards and risks must be developed and carefully reviewed to identify means to enhance process safety, particularly inherent safety.

- *Compliance with Standards* – These include industry standards, regulations and RAGAGEP. There are many design standards based on years of learnings that are available to guide a project design.

- *Hazard Identification and Risk Analysis* – The project should identify hazards and employ inherently safer design approaches as the first step in managing risk across the asset life cycle. Risk analysis aids in decisions on risk reducing measures.

- *Asset Integrity and Reliability* – The design greatly affects this element. Selection of materials appropriate for the design conditions and designing equipment to facilitate maintenance both serve to maintain integrity.

- *Workforce Involvement* – Operations personnel should be included in the design team. This helps to ensure that operational experience is included in design decisions, not just for hazard identification, but to aid operability and maintainability issues, and to contribute to inherently safer design.

- *Management of Change* – Changes to design should be reviewed, especially after hazard identification, in case these impact safety or environmental barriers assumed present in the hazard identification study.

During both greenfield and brownfield construction, and in initial start-up, additional RBPS elements become important including the following.

- *Contractor Management* – This addresses construction safety and SIMOPS control. Many people may work on different activities in a small space, and managing this work is key to safety including verifying competencies, required certifications and that personnel understand the interface requirements.

- *Operational Readiness* – This activity verifies that the facility is complete and ready for operation. Are all the safety features agreed in design now implemented in the constructed facility and is it safe to start-up?

- *Safe Work Practices* – This addresses safety during construction and ongoing operations. *Safe Work Practices* are essential to support the safe working on the project during construction, commissioning, startup, and into production.

- *Operating Procedures* – Procedures should be developed for startup and ongoing operations. Before startup, these procedures need to be written, available and understood.

- *Training and Performance Assurance* – This element aims to ensure that personnel are well trained for startup and for operations. This training should ensure understanding of *Safe Work Practices* and *Operating Procedures*.

7.1.2 Project Life Cycle Terminology

There are multiple terminologies used by different companies for life cycle stages; this chapter uses the terminology from the CCPS *Guidelines for Integrating Process Safety into Engineering Projects*. The main stages are listed in Table 7-1. The life cycle phases for wells typically follow ISO 16530 and similar documents: basis of design, design activity, construction, production, interventions/workovers, and abandonment.

Table 7-1. Project life cycle stages

Project Life Cycle Stage	Activities
FEL-1 (Appraise)	Front End Loading-1 (FEL-1): Develop a broad range of project options, assess commercial viability, and rank feasible options to take forward.
FEL-2 (Select)	Alternative concept options are evaluated, maximizing opportunities and minimizing threats or uncertainties. A single concept is normally chosen at this stage.
FEL-3 (Define)	A basic design is developed with layout, process flow diagrams, material and energy balances, and equipment data sheets. The timescale and costing from FEL-2 is updated. If sufficient confidence exists, the financial investment decision may be made.
Detailed design	Detailed engineering is carried out based on the final option chosen. This includes P&ID drawings, detailed layout, equipment specification, fire protection, etc. suitable to allow commercial bids.
Construction	After the successful bidder is selected, this stage includes fabrication, construction, installation, pre-commissioning, and initial start-up. Offshore fixed structures are usually constructed onshore and floated out to the location, the jacket is installed on the seabed and topsides are crane lifted onto the structure. Floating facilities are constructed in shipyards and towed to their final location.
Start-up	There is a commissioning and start-up phase to ensure that the completed facility meets its design specifications. The timing of handover to the customer varies between companies and projects – some before commissioning, some before start-up, some after meeting design specifications.
Operation	This stage includes project phase out and handover to the operations team. During operations, projects may include small projects and MOC changes. It includes SIMOPS (e.g., as new wells are drilled or older wells reconditioned or remediated during operations).
Abandonment	Facilities must be safely decommissioned, and all equipment removed. Wells must be sealed and abandoned. Offshore fixed facilities are usually cut off below the seabed to avoid future snagging issues, although some jurisdictions allow toppling onto the seabed. Abandonment should be treated like a new project.

A diagram showing the typical project stages is shown in Figure 7-1. Smaller projects and MOCs during the operations phase and end of life abandonment are broken out to show they also follow a formal design approach.

As is shown in Table 7-1 and Figure 7-1, the life cycle is a staged process gradually refining an opportunity with several options (FEL-1) into a selected option based on evaluating economics, safety and environmental issues, and project execution risks (FEL-2). A basic design is carried out (FEL-3) and key assumptions from the earlier stages are verified. A detailed design follows and includes all equipment specifications and drawings, suitable for bidding. Once the successful bid team is selected, and necessary approvals obtained, then construction is initiated, followed by initial start-up. A final handover to the operations team occurs and the facility is managed by them through life, including small projects and MOC activities, until end of life and abandonment. Process safety should be embedded in every stage and this chapter provides a summary of some of the major activities and linkages to RBPS.

In principle, onshore projects and offshore projects follow the same life cycle stages, but often with different specialist contractors to address risks specific to the facility.

7.2   FRONT END LOADING

The first stage activities are termed Front End Loading (FEL). Well construction projects typically occur before the first FEL stage. Decisions to drill a well are based on evaluation of seismic data and data from nearby fields. Once a well or follow-up appraisal wells are drilled to determine reservoir characteristics and a commercial development opportunity is identified, then design of additional wells and production facilities follow the life cycle stages defined in Table 7-1.

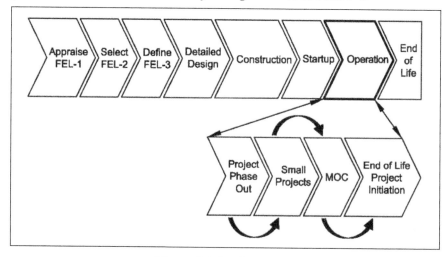

Figure 7-1. Project stages
(CCPS, 2019b)

The three FEL stages carry out similar process safety activities, but with greater refinement as the project option is selected and refined. Some companies develop a process safety in design philosophy to ensure the appropriate process safety activities occur at each stage.

7.2.1 FEL-1

The primary activities in FEL-1 relate to *Compliance with Standards* (particularly the regulatory aspects that must be complied with) and *Hazard Identification and Risk Analysis*. At this stage, there are multiple options but with few details yet developed. This may appear as simple block diagrams of major parts of the process. Decisions address location, technology, process and how to achieve inherently safer design. Standards and prior projects help with some design assumptions, and *HIRA* takes the form of simple checklists of potential hazards (e.g., CCPS, 2019b Table 3.1). The aim at this stage is to rank possible project design options based on likely economic advantages, process safety (including environmental impact), and other factors such as project risk.

It is good practice to develop a risk register at this stage. This risk register lists hazards, their potential consequences, their likelihood, and overall risk. The register is used to support project decisions and is updated as the design progresses. This helps to ensure that no identified risk is neglected when there is a design modification and MOC, or if the project team changes or new personnel join, as happens over the life cycle. The risk register may also identify inherently safer options where some risks may not be present at all, and thus not require mitigation.

Inherently safer design (ISD) activities start in FEL-1 when there is the greatest potential to eliminate or minimize risks. Onshore this might be to choose options with immediate export of produced gas and liquids and without any local storage. Offshore this might include some partial subsea processing or a very low staffing option for operating "from the beach". An example is presented later in this chapter that further illustrates ISD concepts.

Some projects carry out an initial Concept Risk Analysis (CRA) based on assumptions for inventories and process conditions. CRA can be a simplified QRA or a simpler assessment limited to potential consequences of worst case or maximum credible events that uses basic information only. It may use prior similar designs to allow sensible assumptions. At the FEL-1 stage, the CRA is more of a screening level risk estimate, and it is useful for comparing different options or identifying major risks that may be difficult to mitigate at later design stages or in operation.

7.2.2 FEL-2

The primary objective of FEL-2 is to select the option to take forward in the project. As before, the primary activities in FEL-2 relate to *Compliance with Standards* and *Hazard Identification and Risk Analysis* but now sufficient detail is available such that issues related to *Asset Integrity and Reliability* can be defined.

FEL-2 develops selected options further and usually one option is progressed to initial mass and energy balances, outline layout and PFDs, and equipment lists to allow an outline costing to be developed.

Once the final option is selected, the initial *Hazard Identification and Risk Analysis* studies, risk register, Inherently Safer Design review, and Concept Risk Analysis are all updated. Final option selection involves a multi-variable balancing of project finance, potential process safety and environmental impacts, and project risks (e.g., construction risks, weather).

Initial engineering design activities are undertaken including: establishing the design philosophy, identifying relevant regulations and standards, facility siting or module layout, conducting preliminary studies for fire and explosion analysis, fire and gas detection, fire hazards, firewater and foam needs, blowdown and depressurization, transportation risk, and security vulnerability.

The ISD is updated in FEL-2 as part of the final option selection process. Sutton (2011) describes the application of inherent safety into an offshore FEL stage design. Options are screened out based on good design principles and the ISD hierarchy – using Figure 7-2 as a guide for ranking. The ranking is discussed in the box explaining the concept of Inherently Safer Design. ISD options include those related to workforce exposure and the use of reduced personnel levels or even unmanned facilities, thus reducing risks. Other ISD options may relate to facility layout and sizing of hydrocarbon storage. There may be a trade-off where safety is enhanced, but environmental impacts increased (e.g., offshore subsea processing mostly eliminates safety issues, but smaller undetected leaks persist for longer than if on a manned facility). Both aspects are important, and the design selection should consider each in the option selection.

Where a CRA (either QRA or consequence analysis) is performed, it is refined at the FEL-2 stage when outline design activity occurs, and it is part of the ranking of the different options. As before, the risk estimates are at more of a screening level of detail but may be sufficient to differentiate options.

There may be adequate information at this stage to address *Asset Integrity and Reliability*. It might establish the need for more costly alloy materials to deal with corrosive fluids. It might also address whether a single equipment item is sufficient or there should be redundancy to provide increased reliability.

The relationship between various risks and risk reduction options should be considered at this stage. For example, where a normally unmanned installation is proposed which reduces personnel risks, then the equipment should be designed with sufficient integrity to minimize the need for maintenance (which could inadvertently increase manning levels).

Inherently Safer Design

Trevor Kletz was an early proponent of inherent safety and CCPS formalized his papers into books, initially in 1996, again in 2009, and most recently into a Guideline CCPS (2019a). These updates did not change the concepts but refined the implementation examples. Some key words used to describe inherent safety include:

- *Minimize*: Reduce inventories of hazardous materials
- *Substitute*: Replace hazardous material or process with a more benign one
- *Moderate*: Use less hazardous processing or storage conditions
- *Simplify*: Eliminate unneeded complexity and make designs error tolerant

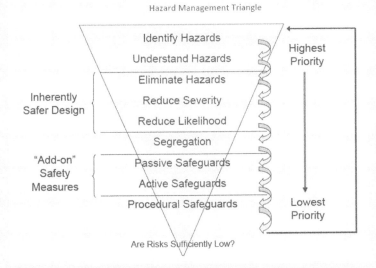

**Figure 7-2. Concept for including inherently safer design
(Broadribb, 2010)**

This figure incorporates ISD in a hierarchy of safety measures – with eliminate hazards being the preferred option but followed by reduced severity and reduced likelihood options. Segregation uses layout to separate hazards and reduce escalation. Passive and active safeguards improve safety for people, but do not address the inherent hazards of the process or the materials, instead mitigating their potential consequences. The final measure, procedural safeguards, which is least reliable, is the lowest category and it relies on personnel actions to reduce risk.

7.2.3 FEL-3

This stage takes the basic design from FEL-2 and refines it further. Earlier *HIRA* studies (including PHA, ISD and CRA) are updated to reflect the greater design detail available to ensure hazards have been identified, inherent safety principles have been applied as much as practicable, and potential residual risks or consequences have been evaluated. Additional activities may be undertaken, such as Layer of Protection Analysis (LOPA) building on the PHA, to establish whether additional risk reduction is required for higher risk scenarios. This may result in additional isolations or functional safety equipment following for example IEC 61511. *HIRA* methodologies are covered in detail in CCPS (2008a) and CCPS (1999). Full details on the LOPA methodology appear in a series of CCPS texts (CCPS 2001, 2013a, 2015).

A full QRA might be carried out at detailed design stage for the more complex facilities or if required by local regulations. Refer to Section 6.3.3 for an outline of QRA content.

Detailed layout evaluations occur at this stage, building on earlier risk studies. Onshore installations usually employ a facility siting approach (API 752 and CCPS, 2018b) to ensure protection of occupied buildings and any affected public. Offshore installations start with broad principles of layout, for example separating the most hazardous items from occupied spaces (control room, accommodation) and siting risers in locations where collision with marine vessels is unlikely, etc. This may be guided by relevant standards (API 14J) or by a QRA to model incident scenarios and their potential escalation. Where escalation cannot easily be prevented by physical separation, then passive barriers such as firewalls or blast walls may be specified complemented by active barriers (fire and gas detection, isolation, depressurization, and blowdown). Exclusion zones around proposed flare(s) also need to be determined.

Additional studies conducted at FEL-3 may include reliability, availability, and maintainability (RAM) assessment; safety critical equipment specification (EI, 2007); safety instrumented level (SIL) assessment; hazardous area classification; evacuation, escape and rescue analysis; human factors analysis; fire and gas detection and shutdown philosophy; relief, blowdown, and flare assessment; dropped object study; and security vulnerability analysis. Details for these studies are included in CCPS (2019b).

Some companies carry out a review at the end of FEL-3 as well as other stages to assure that all aspects of design, including process safety, have been adequately addressed. This is an example of the *Auditing* element of RBPS. A checklist for this review is included in the CCPS (2019b) *Guidelines for Integrating Process Safety into Engineering Projects.* The number of reviews varies between companies/projects, but five or size is typical for major projects. Smaller projects may combine reviews. These reviews are conducted by an independent team of experienced engineers, operators, and other appropriate disciplines.

The output of FEL-3 is a detailed design package, often termed a Basis of Design (BOD) document, with sufficient information to allow commercial bids to be obtained, but also allowing bidders to incorporate their own ideas to enhance efficiency, improve process safety, or reduce project risk.

7.3   DETAILED DESIGN

Many activities carried out in detailed design are a refinement of studies already defined in the earlier front end loading activities, especially FEL-3. RBPS elements: *Compliance with Standards* and *Hazard Identification and Risk Analysis* dominate, but other key activities at this stage include the following (with a single example activity – but many more may be required).

- *Process Safety Culture*
  - o   Build a culture in the design team that recognizes process safety issues and prioritizes these equally to other safety issues
- *Process Safety Competency*
  - o   Train personnel in process safety issues and past lessons learned
- *Workforce Involvement*
  - o   Involve operations and maintenance personnel in design activities and reviews to ensure their knowledge is incorporated, and assist with development of operating procedures and emergency response
- *Process Knowledge Management*
  - o   Document relevant process safety issues – material properties, corrosion, toxicity, flammability etc.
- *Management of Change*
  - o   Changes to the design need to be formally reviewed, especially any changes that might affect HAZID assumptions
- *Operating Procedures*
  - o   These need to be developed as part of the design package
- *Emergency Management*
  - o   The full escape, evacuation and rescue system should be developed

The risk register is an important input to the design team to assure that risks identified by the prior FEL Team and any required actions are not inadvertently neglected or undermined. Similarly, a standards and regulation review is undertaken to ensure nothing has changed since the FEL stage. Certain activities or studies may be required during the design stage to meet local regulations and to obtain a permit to operate in a timely manner.

The *HIRA* studies are repeated or refreshed, but now with full process details, including P&ID, equipment details, the instrument and control system, the F&G detection and shutdown system, depressurization and blowdown details, etc. The *HIRA* team includes the designers and workforce representatives, if available, or experienced personnel from other installations, if not. Consideration should be given to have the leader and perhaps other members independent of the design team.

The earlier section on FEL-3 listed multiple safety studies – all these should be updated with final design details. The process safety studies are integrated with many other safety and environmental studies and these typically overlap FEL-3 and Detailed Design. The *Guidelines for Integrating Process Safety into Engineering*

Projects (CCPS, 2019b) provides advice as to which process safety studies occur when in the design activity.

7.4   PROCUREMENT AND CONSTRUCTION

Once the detailed design is complete, with all associated process safety assessments, then the project goes out for procurement and construction bidding. Important RBPS

Incident: Piper Alpha Design Issue

The Piper Alpha incident has been described previously in Chapter 6. While primarily due to miscommunication related to work permit status that allowed a startup with parts of the pipework still open, an important design issue also existed. The facility had been very successful and highly productive. The operator sought an increase in production, and this was granted, subject to a condition that gas also be piped to shore and not flared.

A gas treatment plant with compression was retrofitted, but the facility design was not changed to add blast walls to account for the greater potential of an explosion event. Also, the compression module was close to occupied spaces (refer to Figure 6-2 in Chapter 6).

Broadribb (2014) discusses important changes to design that occurred due to this incident and the Cullen Inquiry recommendations. Among other things, the Inquiry recommended that, forthwith, all facilities in the UK carry out certain studies to enhance inherently safer design. These are known as the 'forthwith studies'.

- Systematic analysis of fire and explosion hazards
- Analysis of smoke and gas ingress into living quarters, and the requirement for a temporary (safe) refuge capable of surviving the initial fire/explosion and any escalation for a reasonable duration to permit evacuation and escape
- Analysis of the vulnerability of safety critical equipment or elements, such as emergency shutdown valves (ESDVs)
- Analysis of escape, evacuation and rescue in the event of major incidents

One effect of the forthwith studies was that the design of North Sea installations changed from essentially square cross sections to rectangular. This allowed for increased spacing between major hazard modules and vulnerable areas. Where feasible, bridge linked platforms were used to further increase separation to the accommodation module.

RBPS Application

Management of Change and *Hazard Identification and Risk Analysis*: The changes recommended by the Cullen Inquiry are now commonly applied to offshore installation design process safety studies and are embedded in UK regulations.

---

elements for this stage include: *Compliance with Standards* (including local regulatory requirements), *Safe Work Practices* (to address construction safety), *Contractor Management* (to assure that personnel assembling items are well qualified and understand safety requirements), *Operational Readiness* (to confirm the constructed design is ready to be safely operated), and *Management of Change* (to assess any change from the design specification).

7.4.1 Procurement and Quality Plan

Some long lead time items, such as large compressors, may need to be ordered and procured before detailed design is complete. For these, the full range of process safety assessments related to them, as listed in FEL-3, need to be completed earlier; changes after ordering are difficult and expensive.

A quality plan should be established to ensure that procured items for well construction and production facilities match the specifications (e.g., material, grade, and documentation trail). The plan should address the roles and responsibilities for quality assurance (QA) and quality control (QC) between the owner, contractors, and suppliers. Having a plan to control quality is key to future *Asset Integrity and Reliability* and underpins process safety.

Equipment and materials delivered to the worksite should be logged. Alloys should be specially marked, stored in separate areas, and use a positive material identification program to ensure that normal steels and alloys are not mixed.

QA/QC procedures related to equipment to be assembled is important. Critical equipment may be subject to a witnessed Factory Acceptance Test (FAT) before delivery. There is a growing awareness of the importance of factory designed software impacting process safety. Special software reviews are now being undertaken using the software functional design specification to assure that all inputs are handled correctly. Contracts with vendors need to allow for such reviews.

7.4.2 Construction

Construction of upstream facilities is different between onshore and offshore. Onshore construction is often onsite with material storage areas and fabrication adjacent to the final installation. Offshore construction of topsides is usually undertaken in marine yards and transported to the final location. For fixed platforms, large marine cranes are used to lift the fully or partially fabricated topside modules onto the structure. For deepwater, FPSOs and FPUs are usually fully constructed in the yard and towed (or sailed) to the final location where risers and other connections to the wellhead are made. Some remote locations (e.g., Australia Northwest Shelf) use offshore techniques of yard construction and towing to the site for final assembly of modules even for onshore facilities.

Safety during construction is primarily related to occupational safety as hydrocarbons are typically not yet introduced and thus process safety risks are minimal. An important exception is SIMOPS, where brownfield construction activity is occurring adjacent to already operating facilities. Permit issue trailers and weather shelters should be sited appropriately in operating facilities (API RP 753 for temporary facilities). This was discussed in Chapter 5. Because construction activities (e.g., lifting or welding) increase potential risks from the operating facility, these must be managed effectively.

7.4.3 Operational Readiness

Once construction is completed, *Operational Readiness* is ascertained through a full verification that the installation meets the design specification, including process safety design aspects, and that actions identified have been tracked to close-out. A punch-list of carry-forward actions is developed to complete before, during, and after commissioning. The compilation of a full documentation set usually coincides with the conclusion of construction. The full list of documents is extensive and is provided in CCPS (2019b). More important process safety documents include: the risk register, commissioning procedures, operating procedures, safe work practices, emergency response plans, punch lists and action tracking. A review, independent of the construction team, is often undertaken to verify that all safety, environmental and technical risks, are well managed.

7.5   COMMISSIONING AND STARTUP OF FACILITIES

Commissioning is the process of assuring that all the systems and equipment are tested and safe to operate. This is included in the RBPS element *Operational Readiness.* Pre-startup safety review is one of the SEMS requirements and also applies in goal-based regulations. Startup is the process of introducing hydrocarbons or other hazardous chemicals to the facility to establish operation.

Pre-commissioning activities may be undertaken to test individual systems before the entire facility is completed. Pre-commissioning may be treated as a SIMOPS activity and managed using *Safe Work Practices* (see Chapter 5) to avoid potential interactions. CCPS (2019b) covers many of the details.

Before startup offshore, there are normally specific local regulatory approvals required before hydrocarbons are introduced. Onshore approvals are different and typically determined in the US at state or county level except for larger facilities that must comply with OSHA PSM and EPA RMP regulations. Local approvals range from only a few to very extensive requirements depending on local sensitivities. Normally this involves addressing *Emergency Management* by contacting local emergency first responders as they supplement the facility's own personnel. Onshore facilities may additionally wish to engage *Stakeholder Outreach* by contacting neighbors to foster positive relations going forward or this may be organized by the local authority.

Training and Performance Assurance of future operational personnel should be part of the commissioning and startup activity and this training normally includes a mix of designers, the construction team and the company's own training and process safety personnel. Training focuses on *Operating Procedures* and *Safe Work Practices.*

8

Process Safety: Looking Forward

8.1   LOOKING FORWARD

The previous chapters have focused on design, construction, and operation of today's onshore and offshore oil and gas facilities with illustrations based on historical incidents. The future will undoubtedly be subject to change. A key focus area will be to build and maintain robust and resilient organizations that have multiple means to actively determine the risk the organization creates in its operations and builds active protections against those risks. That change will be prompted by research, by advances in technology, and by influences from the surrounding business and world.

Once change is agreed and made, then other needs can be addressed in a process of continual improvement. Thus, what is stated here may be subject to revision sooner than the material in earlier chapters. This is neatly summed up in the following quote.

Remember, in two days' time, tomorrow becomes yesterday.

Looking forward includes developing the vision for process safety for upstream. This chapter identifies some current research needs and indicates an approach for developing a future vision for process safety.

There is no direct RBPS Element focusing on looking forward, but all four RBPS pillars have issues that would benefit from further development.

- *Pillar: Commit to process safety*
- *Pillar: Understand hazards and risks*
- *Pillar: Manage risk*
- *Pillar: Learn from experience*

Also, the field of process safety is not isolated from other socio-technical developments. It will be impacted by change due to the following.

- Digitalization and large data analytics
- Climate change measures
- Sustainability
- Cyber threats
- Culture and organization

The following sections summarize some research needs identified by regulators and by research bodies and includes discussion on selected technical advances.

8.2   RESEARCH NEEDS

In the following sections, broad areas for research are identified by various bodies. If readers wish further detail on these areas of research, they can refer to the source documents, most of which are freely available on the web.

8.2.1 Regulatory Bodies

In the US, OSHA sponsors research related to onshore facilities, but these relate more to occupational safety than to process safety. The EPA sponsors gas dispersion trials (e.g., Desert Tortoise and Jack Rabbit trials) to measure and facilitate modeling of gas dispersion over terrain and around obstacles using wind tunnels and outdoor experiments. The Chemical Safety Board makes many recommendations following process safety incidents, including a few related to upstream facilities, but does no research itself.

NFPA, FERC (Federal Energy Regulatory Commission) and HSL (UK Health and Safety Laboratory) collaborated over several years to develop a model evaluation protocol to verify potential dense gas dispersion models for LNG facility approvals. Kohout (2012) reported that three models tested against the protocol were approved (PHAST, FLACS and DEGADIS). While specific for LNG releases, the protocol was based on good dispersion science, and this work supports analysis of potential releases from onshore upstream facilities.

The UK HSE has defined its research needs and achievements (http://www.hse.gov.uk/research/rrhtm/index.htm). Most of these are technical projects, addressing a range of occupational safety and process safety topics, but occasionally the HSE issues more philosophical papers related to process safety. These tend to be strategic papers setting out how the HSE assesses process safety issues (e.g., Reducing Risks Protecting People originally issued in 2001). An important process safety topic for the HSE is how best to deploy risk assessment and how to achieve worker involvement. The UK HSE also sponsored research into the Buncefield vapor cloud explosion incident in 2006. While the Buncefield terminal is not an upstream facility, the learnings related to vapor cloud explosions and potential escalation to detonations do apply to upstream.

There are multiple regulatory body initiatives documenting research needs for process safety. The International Committee on Regulatory Authority Research and Development (ICRARD) helps to coordinate these various initiatives to encourage cooperation and to prevent duplication.

BSEE has a program to evaluate safety-related technologies, the Technology Assessment Program (TAP). This has administered nearly 900 research and development projects since its inception. The majority of these are detailed technical topics rather than broader process safety topics, although many address risks that could potentially result in loss of containment incidents. One project examined how the SEMS regulation might be expanded to more explicitly incorporate process safety topics such as leading and lagging indicators, human factors, ALARP, barrier performance and continuous improvement (ABSG, 2015). While not currently adopted by regulation, BSEE has supported the revised fourth Edition of API RP 75 which allows greater freedom to operators to design a safety and environmental management system to incorporate enhanced process safety concepts.

8.2.2 Research Organizations

There are multiple organizations focusing on research needs, both onshore and offshore. Much of the onshore process safety research is generally applicable to any onshore facility and is not focused towards upstream. Onshore research related to fire and vapor cloud explosions is carried out by multiple organizations such as the UK HSE Health and Safety Laboratory, NFPA, DNV GL Spadeadam, and by BakerRisk for the Explosion Research Cooperative, and often reported through FABIG (a group within the Steel Construction Institute). The Mary Kay O'Connor Process Safety Center (MKO) has established an LNG experimental facility and they have investigated fire response and fire-fighting approaches. MKO also carries out process safety research on a wide variety of topics including advanced risk assessment methods, fire detection, operating procedures effectiveness, and chemical reactivity. A consortium of companies led by Honeywell supports Abnormal Situation Management – the ASM Consortium. The US DOE Sandia Laboratory carried out the largest LNG fire experiment on water and thus helped siting of LNG facilities. NFPA carries out research across the full range of fire hazards and some of these relate directly to process safety.

In the US, some important organizations for offshore oil and gas development include COS (Center for Offshore Safety), Subsea Systems Institute, RPSEA (Research Partnership to Secure Energy for America), the Society of Petroleum Engineers SPE-GRP, and OESI (Ocean Energy Safety Institute). OESI has issued a research roadmap (OESI, 2019).

The OESI report notes that four disciplines underpin offshore safety – human factors, process safety, SEMS, and technology. The same is true for upstream onshore safety as well, although the technology may differ in some respects. Sharing of incident information helps industry *Learn from Experience* and improve process safety performance. OESI notes the benefits of sharing data from SafeOCS on a loss of well control event that was addressed better because of this information. COS shares information on findings from specific incidents and high value learning events and from SEMS audits to encourage continual improvement by all companies. There is less industry-wide data collection and sharing for onshore upstream.

The National Academy of Science, Engineering and Medicine – Gulf Research Program (GRP) worked with the Society of Petroleum Engineers to sponsor a summit in 2018 to review safety and environmental research needs and to address major incident prevention (SPE, 2018). This summit identified 136 opportunities for potential research. Some key forward needs developed in the summit included the following.

- Improve collaboration between industry, regulators, and academia
- Spur innovation aimed at reducing or managing risks

- Create robust and resilient organizations
- Identify educational programs promoting a skilled and safety-oriented workforce

A follow-up SPE meeting in 2018 also identified a need to change some mindsets. This reaffirmed the above topics but also included changing from a "goal" of zero to an "expectation", greater use of leading metrics, and to improve application of human factors.

RPSEA (2018) issued a roadmap for R&D needs addressing onshore, offshore and environmentally sensitive/arctic areas. The RPSEA recommendations related to process safety were more technical than those listed above and addressed topics such as: next generation well control and BOPs, real-time downhole detection of hydrocarbons during drilling operations, and improvements in life of the well integrity, cementing, and casing.

The OESI (2019) research roadmap integrated the above reports and developed an updated list. While focused on offshore, many of the topics are of relevance to onshore.

Norway carries out extensive research related to upstream process safety, often carried out by SINTEF and by universities at Trondheim and Stavanger.

Classification Societies also carry out extensive research for upstream offshore topics, usually related to detailed technical matters. DNV GL operates the Spadeadam test facility which has carried out hundreds of large-scale vapor cloud explosion trials, mostly deflagrations. It is currently working on a research program, triggered by the Buncefield incident, on the potential for foliage to lead to even more serious DDT (deflagration to detonation transition) events. This is relevant to onshore upstream production facilities.

8.3   TECHNICAL ADVANCES

There are many technical changes that may be expected to advance in the coming years.

8.3.1 Process Safety

While some of the research programs identified may lead to strategic change, at a lower incremental level all the process safety tools described in earlier chapters of this book are likely to receive improvements. Some more significant improvements might be expected in the following areas.

- Increased focus on and development of tools supporting the integration of human factors into the workplace, including the equipment we work with, the way we are organized, and the processes that define our work
- New and improved ways to engage the workforce and create stronger safety culture

- Tighter integration of hazard identification, risk assessment, and risk registers over the complete facility life cycle
- Improved risk assessment tools (e.g., LOPA customized for upstream, QRA, and quantification of bow tie analysis)
- Improved mechanical integrity programs for safety critical items and techniques to determine current barrier status
- Improvements to well integrity and kick detection
- More robust likelihood estimates and tools (e.g., onshore failure datasets, Bayesian analysis)
- Use of digitalization: to extract information and lessons learned, data centric condition monitoring, predictive analytics to pre-empt failure and initiate repair, replacement or maintenance, etc.

8.3.2 World Energy Source Transition

As the world seeks and moves to more environmentally sustainable energy sources, there will be a reduced dependence on oil. This will impact upstream process safety in a number of ways.

- There will be an increase in new technologies to produce oil more cost-effectively. Introduction of new technologies means the potential introduction of new hazards. These should be managed through *Hazard Identification and Risk Analysis*. As we have seen in the past, the challenge for more cost-effective production often comes with a challenge to topics such as *Asset Integrity and Reliability* and *Training and Performance Assurance* that will need to be addressed.
- As there is a reduction in dependence on oil, there will be an increase in gas production. As was described in the short primer on vapor cloud explosions in Section 5.3.2, gas handling will increase risks, which will necessitate changes in facility design and operations.
- Digitalization will modernize many of the workflows in upstream. Work will occur at a faster pace, sometimes in parallel, and increasingly using artificial intelligence. This will create a challenge to the implementation of many of the RBPS elements including *Operating Procedures*, *Safe Work Practices*, and *Management of Change* to name a few. They will need to happen more quickly, incorporate learnings more quickly, and be more robust. The topic of human performance will be key as the operator and maintenance technician day-to-day job tasks will change.
- Integrated value chains will change business sector relationships. A growing hydrogen economy will introduce new risks. Carbon capture and sequestration may prompt a reduction in decommissioning of installations and an increase in repurposing these installations to facilitate sequestration.

8.4 VISION FOR UPSTREAM PROCESS SAFETY

The Guidelines for Risk Based Process Safety set out the content of a comprehensive process safety management system, but this alone is unlikely to be sufficient. As CCPS found for the downstream industry, a forward vision is a necessary complement to RBPS. Grounds et al (2015) describe Vision 20/20 as looking into the not-too-distant future to describe how great process safety is delivered when it is collectively and fervently supported by industry, regulators, academia, and the community worldwide.

Vision 20/20 is not simply a vision statement; it defines tenets and provides a self-assessment tool. At a high level, tenets include: a committed culture, vibrant management systems, disciplined adherence to standards, intentional competency development, and enhanced application and sharing of lessons learned. Initial application of the self-assessment tool showed that even for companies following RBPS, there were real gaps (Grounds, op cit). Knowledge of the gaps then becomes a starting point for improvement.

CCPS and SPE believe that a similar vision on process safety based on these high-level tenets will ultimately benefit both the upstream and downstream industries. This vision would assist the industry and all of its stakeholders achieve a superior level of process safety more quickly.

In closing, it goes without saying that leadership is fundamental in supporting strong process safety performance in upstream, as in any other industry. Leadership at all levels should strive to design, operate, and maintain a workplace where all workers support each other in conducting work successfully and safely – in other words, creating a strong process safety culture.

REFERENCES

ABB (2013) Oil and gas production handbook, 3rd Ed, Editor Håvard Devold, available from ABB website: http://www04.abb.com/global/seitp/seitp202.nsf/0/f8414ee6c6813f5548257c14001 f11f2/$file/Oil+and+gas+production+handbook.pdf

ABSG (2014) Analysis of Domestic and Foreign Oil and Gas Standards (Final Report), available from BSEE https://www.bsee.gov/research-record.

ABSG (2015) Process safety Assessment – Final report, submitted to BSEE. Available on BSEE website https://www.bsee.gov/research-record/tap-732-process-safety-assessment.

API (1990) Management of process hazards. American Petroleum Institute, API RP 750, Washington DC.

API (2001) Recommended practice for design and hazards analysis for offshore production facilities, API RP 14J, 2nd Ed., reaffirmed 2019, American Petroleum Institute, Washington.

API (2006) Recommended practice for design of offshore facilities against fire and blast loading, API RP 2FB, American Petroleum Institute, Washington.

API (2007) Management of Hazards Associated With Location of Process Plant Portable Buildings, API RP 753, 1st Ed, American Petroleum Institute, Washington, DC.

API (2009) Management of Hazards Associated With Location of Process Plant Permanent Buildings, API RP 752, 3rd Ed, American Petroleum Institute, Washington, DC.

API (2010) Isolating Potential Flow Zones During Well Construction, API 65-2, 2nd Ed, American Petroleum Institute, Washington DC.

API (2012) Fire Protection in Refineries, API RP 2001, 9th Ed., American Petroleum Institute, Washington DC.

API (2013a) Deepwater Well Design and Construction, API RP 96, First Edition, American Petroleum Institute, Washington DC.

API (2013b) Well Construction Interface Document Guidelines, API Bulletin 97, American Petroleum Institute, Washington DC.

API (2015a) API Standards for Safe Onshore Operations and API Standards for Safe Offshore operations, lists available from API website: https://www.api.org/oil-and-natural-gas/health-and-safety/exploration-and-production-safety

API (2015b) Hydraulic fracturing – well integrity and fracture containment, API RP 100-1, American petroleum Institute, Washington DC.

API (2015c), Choke and Kill Equipment, API Spec 16C, 2nd Edition, Washington DC.

API (2017) Recommended practice for analysis, design, installation, and testing of basic surface safety systems for offshore production platforms, API RP 14C, 8th Ed., American Petroleum Institute, Washington.

API (2018) Well Control Equipment Systems for Drilling Wells, API 53, 5th Ed, Dec 2018, American Petroleum Institute Washington DC.

API (2019a) API Recommended Practice 75 – Recommended Practice For Development Of A Safety And Environmental Management Program For Offshore Operations And Facilities, API RP 75, 4th Ed, American Petroleum Institute, Washington DC.

API (2019b) Spec 5CT/ISO 11960: Specification for Casing and Tubing, 10th Ed, American Petroleum Institute, Washington DC.

Baker Panel (2007) Report of the BP U.S. refineries independent safety review panel, Chairman James Baker.

Barusco, P. (2002) The Accident of P-36 FPS, Offshore Technology Conference, OTC 14159, Houston, May.

Broadribb, Michael (2010) HAZOP/LOPA/SIL Be Careful what you Ask for!, 6th Global Congress on Process Safety.

Broadribb, Michael (2014) What have we Really Learned? Twenty Five Years after Piper Alpha, Process Safety Progress, Vol. 34, No. 1.

Broadribb, Michael (2017) And now for Something Completely Different, Process 13th Global Congress on Process Safety.

BSEE (2013) Investigation of November 16, 2012, Explosion, Fire and Fatalities at West Delta Block 32 Platform E, BSEE Panel Report 2013-002, Bureau of Safety and Environmental Enforcement, Washington.

BSEE (2016) Final Drilling Rule, 30 CFR 250, Federal Register, Vol 81, No. 83, April 29.

BSEE (undated) Final Safety Culture Policy Statement, available at www.bsee.gov

CCPS (1991) Technical Management of Process Safety, AIChE New York. Now superseded by CCPS Risk Based Process Safety.

CCPS (1995) Guidelines for Technical Planning for On-Site Emergencies, Wiley/AIChE, New York.

CCPS (1999) Guidelines for chemical process quantitative risk analysis, 2nd Ed., Wiley / AIChE, New York.

CCPS (2001) Layer of Protection Analysis: Simplified Process Risk Assessment, 1st Ed., Wiley / AIChE, New York.

CCPS (2003) Guidelines for fire protection in chemical, petrochemical, and hydrocarbon processing facilities, Wiley / AIChE, New York.

CCPS (2004) Guidelines for Preventing Human Error in Process Safety, Wiley/AIChE, New York.

CCPS (2006) Business Case for Process Safety, 2nd Ed, AIChE website, https://www.aiche.org/sites/default/files/docs/pages/ccpsbuscase2nded-120604133414-phpapp02.pdf

CCPS (2007a) Risk Based Process Safety, Wiley / AIChE, New York.

CCPS (2007b) Human Factors Methods for Improving Performance in the Process Industries, Wiley/AIChE, New York.

CCPS (2008a) Guidelines for Hazard evaluation procedures, 3rd Ed, Wiley / AIChE, New York.

CCPS (2008b) Incidents that define process safety, 1st Ed, Wiley / AIChE, New York.

CCPS (2009) Guidelines for Developing Quantitative Safety Risk Criteria, 1st Ed, Wiley / AIChE, New York.

CCPS (2011) Conduct of Operations and Operational Discipline: For Improving Process Safety in Industry, Wiley / AIChE, New York.

CCPS (2012) Guidelines for Evaluating Process Plant Buildings for External Explosions, Fires, and Toxic Releases 2nd Edition, Wiley / AIChE, New York.

CCPS (2013a) Guidelines for Enabling Conditions and Conditional Modifiers in Layer of Protection Analysis 1st Ed., Wiley / AIChE, New York.

CCPS (2013b) Guidelines for Managing Process Safety Risks During Organizational Change, Wiley / AIChE, New York.

CCPS (2015) Guidelines for Initiating Events and Independent Protection Layers in Layer of Protection Analysis 1st Ed., Wiley / AIChE, New York.

CCPS (2016) Guidelines for Asset Integrity Management 1st Edition, Wiley / AIChE, New York.

CCPS (2018a) Guidelines for Asset Integrity Management 1st Edition, Wiley / AIChE, New York.

CCPS (2018b) Guidelines for layout and siting of facilities. 2nd Ed. Wiley / AIChE, New York.

CCPS (2018c) Bow ties in risk management, joint publication of Center for Chemical Process Safety and Energy Institute, Wiley / AIChE, New York.

CCPS (2018d) Business Case for Process Safety, 2nd Ed, AIChE website, https://www.aiche.org/sites/default/files/docs/pages/ccpsbuscase2nded-120604133414-phpapp02.pdf

CCPS (2018e) Recognizing and Responding to Normalization of Deviance, Wiley / AIChE, New York.

CCPS (2019a) Guidelines for Inherently Safer Chemical Processes: A Life Cycle Approach, 3rd Ed., Wiley/AIChE, New York.

CCPS (2019b) Guidelines for Integrating Process Safety into Engineering Projects, Wiley / AIChE, New York.

CCPS (2019c) More Incidents That Define Process Safety, Wiley / AIChE, New York.

COS (2014) COS SEMS II Audit Protocol- Checklist COS-1-01, COS Website https://www.centerforoffshoresafety.org.

COS (2018) Guidelines for a robust safety culture, COS-3-04, 1st Ed., Center for Offshore Safety, Houston. Available at https://www.centerforoffshoresafety.org/

COS (2020) Annual Performance Report for 2019. Center for Offshore Safety, Houston. Available at https://www.centerforoffshoresafety.org/

CSB (2000) Investigation Report: Catastrophic vessel over-pressurization, Report 1998-002-I-LA, US Chemical Safety and Hazard Investigation Board, Washington DC.

CSB (2019) Investigation Report: Gas Well Blowout and Fire at Pryor Trust Well 1H-9, US Chemical Safety and Hazard Investigation Board, Washington DC.

Crowl D and Louvar J (2019) Chemical process safety, 4th Ed., Prentice Hall.

Deepwater Horizon National Commission (2011) BP Deepwater Horizon Oil Spill and Offshore Drilling – Main Report (Jan 2011) and Chief Counsel's Report (Feb 2011)

Deming, W. Edwards (1993) The New Economics for Industry, Government, and Education, MIT Press, Boston, Ma.

DNV (2011) Forensic Examination of Deepwater Horizon Blowout Preventer, Final Report Vols 1 and 2 for US Dept. of the Interior, Report No. EP030842, Washington DC.

DoE (undated) Research needs for Unconventional Oil and Natural Gas, Washington DC, https://www.energy.gov/fe/science-innovation/oil-gas-research/shale-gas-rd.

Dusseault M, Gray M, Nawrocki P. (2000) Why Oilwells Leak: Cement Behavior and Long-Term Consequences, SPE 64733, SPE Conference Beijing China.

EI (2007) Guidelines for the Management of Safety Critical Elements, 2nd Ed., Energy Institute, London, UK.

EPA (2009) Risk management program guidance for offsite consequence analysis, USA Environmental Protection Agency website,

Franchi J. and Christiansen R. (2016) Introduction to Petroleum Engineering, Wiley, Hoboken, New Jersey.

Grounds C., McCavit J., Scruggs S. (2015) Vision 20/20 with Industry Benchmarking, Global Congress on Process Safety, AIChE available at https://www.aiche.org/ccps/resources/vision-2020.

Halim Z, Janardanan S, Flechas T, Mannan S (2018), In search of causes behind offshore incidents: Fire in offshore oil and gas. J Loss Prev in Proc Ind, 54, 254-265.

Halliburton (2015) Drill well on paper, company website, https://www.halliburton.com/en-US/ps/project-management/well-control-prevention/drill-well-on-paper.html

Hansen O R and Johnson M (2015) Improved far-field blast predictions from fast deflagrations, DDTs and detonations of vapour clouds using FLACS CFD, JLPPI, 35, 293-306.

HSE (1990), Public Inquiry into Piper Alpha Disaster, www.hse.gov.uk › offshore › piper-alpha-public-inquiry-volume1

HSE (1999) Reducing Error and Influencing Behaviour, Publication HSG48, available at https://www.hse.gov.uk/pubns/books/hsg48.htm.

HSE (2009) Provision of active fire protection on offshore installations, Information Sheet No. 5/2009, Health and Safety Executive, UK.

HSE (2013) Managing for Health and Safety (HSG65), available from HSE website, www.hse.gov.uk.

HSE (2019a) Offshore topics, http://www.hse.gov.uk/offshore/topics.htm

HSE (2019b) Reducing risk, protecting people. Updated in 2019. Available on HSE website http://www.hse.gov.uk/risk/theory/r2p2.htm

HSE (website) (https://www.hse.gov.uk/seveso/introduction.htm)

https://www.epa.gov/rmp/rmp-guidance-offsite-consequence-analysis

Hudson (2007) Implementing a safety culture in a major multi-national, Safety Science, 45, 697-722.

IADC (2009) Health, safety and environment guidelines for land drilling contractors, Issue 1.0.1, International Association of Drilling Contractors, Houston.

IADC (2015a) IADC Drilling Manual, Vols 1 and 2, 12th Ed, International Association of Drilling Contractors, Houston.

IADC (2015b) Health, safety and environment guidelines for mobile offshore drilling units, Issue 3.6, International Association of Drilling Contractors, Houston.

IADC (2015c) Deepwater Well Control Guidelines 2nd Ed., International Association of Drilling Contractors, Houston.

IChemE (1996) Risk assessment in the process industries, 2nd Ed., Institution of Chemical Engineers, Rugby, UK.

IChemE (2011) Chemical and Process Plant Commissioning Handbook, 1st Ed., Rugby, UK.

IChemE (2018) Piper Alpha Disaster – 30th Anniversary, Special issue Loss Prevention Bulletin, Rugby, UK.

International Council of Chemical Associations website (https://www.icca-chem.org/responsible-care/)

International Energy Agency (2018) Offshore Energy Outlook, available from www.iea.org, Paris.

IOGP (2018a) IOGP Report 456 – Process safety – recommended practice on key performance indicators, v2. International Association of Oil & Gas Producers, London, IOGP Bookstore and Website (iogp.org)

IOGP (2018b) Safety performance indicators – 2017 data. Available from IOGP Bookshop and Website (iogp.org)

IOGP (2019a) Safety performance indicators – Process safety events – 2018 data, IOGP Bookstore and website (iogp.org)

IOGP (2019b) Risk Assessment Data Directory - Report 434-01 Process release frequencies, https://www.iogp.org/bookstore/product/434-00-risk-assessment-data-directory-overview/

ISO 17776 (2016) Petroleum and natural gas industries -- Offshore production installations -- Major accident hazard management during the design of new installations. International Organization for Standards, Geneva, Switzerland. "©ISO. This material is reproduced from ISO 17776:2000, with permission of the American National Standards Institute (ANSI) on behalf of the International Organization for Standardization. All rights reserved."

ISO 45001 (2018) Occupational Health and Safety Management Systems – with guidance for use, International Organization for Standards, Geneva, Switzerland.

Kohout A, (2012) Evaluation of Dispersion Models for LNG Siting Applications, AIChE Global Congress on Process Safety.

Lacoursiere J-P., Dastous P-A., Lacoursiere S. (2015) Lac-Megantic Accident: What We Learned, Process Safety Progress, 34, 1, 15pgs.

Lees (2012) Lees' Loss Prevention in the Process Industries: Hazard Identification, Assessment and Control, 4th Ed, Butterworth-Heinemann, ISBN-13: 978-013971890.

Longford Royal Commission (1999) The Esso Longford Gas Plant Accident, Report of the Longford Royal Commission, Government Printer, State of Victoria (available on Internet).

Marsh (2020) The 100 Largest Losses 1974-2019, Large Property Damage Losses in the Hydrocarbon Industry, 25th edition. https://www.marsh.com/qa/en/insights/research-briefings/100-largest-losses-in-the-hydrocarbon-industry.html

McLeod F. and Richardson S. (2018) Piper Alpha: The Disaster in Detail, The Chemical Engineer, issue 925/926, IChemE, Rugby UK.

Military Standard 1629A (1980) Procedures for performing a failure mode, effects and criticality analysis, US Dept. of Defense, Washington DC.

NASA (2002) Fault tree handbook with aerospace applications. Rev 1.1, NASA Office of Safety and Mission Assurance, Washington DC.

Norsk O&G (2012) An introduction to well integrity, joint with NTNU and UiS, Norway.

Norsk O&G (2016) Recommended Guidelines For Well Integrity, Document 117, Norwegian Oil and Gas Association, Stavanger, Norway.

NORSOK D-010 (2013) Well integrity in drilling and well operations, Rev 4, Standards Norway.

NORSOK S-001 (2018) Technical safety, 5th Ed, Standards Norway

NORSOK Z-013 (2010) Risk and emergency preparedness assessment, 3rd Ed, Standards Norway.

OESI (2019) 21st Century Ocean Energy Safety Research Roadmap, Ocean Energy Safety Institute, Texas A&M University, College Station, TX.

Oil & Gas UK (2013), Piper 25 Conference, oilandgasuk.co.uk › piper-25-conference.

Oil & Gas UK (2018) Fire and Explosion Guidance, Issue 2, UK Oil and Gas Industry Association Limited, London. and Chapter 9: Risk management plan (Part 68, Subpart G).

Pickles J. and Bain B. (2015) Improved Visualizations of Offshore QRAs, Hazards 25 Conference, IChemE Symposium Series 25, Rugby, UK.

Pitblado R., Fisher M., Nelson B., Fløtaker H., Molazemi, K., Stokke A. (2016), Concepts for Dynamic Barrier Management, JLPPI, Vol 43, 741-746.

PMI (2013) A guide to the project management body of knowledge (PMBOK), Project Management Institute, 5th Ed., Newtown Square, PA.

PSA (2005) Investigation of gas blowout on Snorre A, Well 34/7-P31A, 28 November 2004, Norway Petroleum Safety Authority, Stavanger.

RPSEA (2018) RPSEA Projects, https://rpsea.org/.

SAChE (2019) Safety and Chemical Engineering Education Certificate Program, https://www.aiche.org/ccps/education/safety-and-chemical-engineering-education-sache-certificate-program

Shell (2015) Shell Chukchi Sea Regional Exploration Program Oil Spill Response Plan, rev 3. Public submission available on BSEE website https://www.bsee.gov/sites/bsee.gov/files/research-guidance-manuals-or-best-practices/safety/2015-05-15-revision-3-redacted-shell-chukchi-sea-osrp-with-cover-letters.pdf

Smolen, Brad (2019) Center for Offshore Safety's Annual Performance Report for the 2019 Reporting Year, 7th Annual COS Forum.

SPE (2007) Petroleum Engineering Handbook Vol 2: Drilling Engineering, Editor R. F. Mitchell, 763 pgs, Chapter 9 R. Crook. Society of Petroleum Engineers, Richardson, TX.

SPE (2011) Fundamentals of Drilling Engineering, Eds R.F. Mitchell and S. Z. Miska, 696pgs, Society of Petroleum Engineers, Richardson, TX.

SPE (2018) SPE Summit: Safer Offshore Energy Systems, Society of Petroleum Engineers, San Antonio, TX, May 2018.

Spouge J., (1999) A guide to quantitative risk assessment for offshore installations. Original publisher CMPT

Sutton I. (2011) Inherent Safety in Front End Engineering, 7th Global Congress on Process Safety, AIChE, New York.

Transportation Safety Board of Canada (2014), Lac-Mégantic runaway train and derailment investigation summary.

US National Academies (2016) Strengthening the Safety Culture of the Offshore Oil and Gas Industry, Transportation Research Board Special Report 321, Washington DC.

van Wingerden, K. (2013) Advances in explosion modelling, Piper 25 Conference, oilandgasuk.co.uk webpage, Piper-25-conference.

Vinnem J-E and Roed W (2020) Offshore Risk Assessment, Vol 1 – Principles, Modeling and Applications of QRA Studies, 4th Ed, Springer, London.

Vrålstad T., Saasen A., Fjær E., Øia T., Jan David Ytrehus J.D., Khalifeh M. (2019), Plug & abandonment of offshore wells: Ensuring long-term well integrity and cost-efficiency, Journal of Petroleum Science and Engineering, Vol 173, 478-491.

Wichert E. (2005) Petroleum Engineering Handbook, Chapter 5 Gas treating and processing. Society of Petroleum Engineers.

INDEX